T0271871

Adult Stem Cell Standardization

TECHNOLOGY IN BIOLOGY AND MEDICINE

Volume 1

Series Editor

PAOLO DI NARDO
University of Rome "Tor Vergata"
Rome, Italy

In the past innovation may have been the prerogative of large, closed research labs, but their advantage over smaller rivals and the developing world is now being eroded by two powerful forces. The first is globalization, as more emerging countries become both consumers and, increasingly, suppliers of innovative products and services. The second is the rapid advance of information technologies, which are spreading far beyond the internet and into older industries such as steel, aerospace and car manufacturing. This development is being felt also in the bio-medical domain. The convergence between bio-medicine and engineering holds promises to be highly beneficial for both patients and industries. Words like information technology, advanced materials, imaging, nanotechnology and sophisticated modeling and simulation are now usual in biomedical research and clinical centers. Some scientists and analysts believe that the transformation of biology into an information science instead of a discovery science could be one of the most important innovations in the history of this science, if we are able to activate the untapped ingenuity of people and to accept the risk and the possible failure as central factor of innovation.

The info-medical revolution will be led by information technologies through the establishment of an intelligent network that will enable many other big technological changes. The progressive increase of degenerative diseases and related disabilities together with the extraordinary expansion of longevity will determine the impossibility to sustain health system overspending in most countries. Info-medicine can enhance the quality and quantity of health care, creating equal access to the most advanced diagnostic and treatment procedures by all individuals independently of geographic and economic conditions. The adoption of these technologies, however, depends on absolute proof that they will produce better outcomes and offer value for money.

It is TBM's ambition to represent an international forum to discuss in-depth the most advanced concepts and solutions merging biology, medicine and engineering. To this end, the journal will give scientists the opportunity to publish their most important results reaching the widest audience, while the book series will thoroughly dissect the new concepts in order to supply students and professionals with the most advanced tools to apply them, all in the shortest possible timeframe.

For a list of other books in this series, please visit www.riverpublishers.com

Adult Stem Cell Standardization

Editor

Paolo Di Nardo

University of Rome "Tor Vergata", Rome, Italy

LONDON AND NEW YORK

Published 2011 by River Publishers
River Publishers
Alsbjergvej 10, 9260 Gistrup, Denmark
www.riverpublishers.com

Distributed exclusively by Routledge
4 Park Square, Milton Park, Abingdon, Oxon OX14 4RN
605 Third Avenue, New York, NY 10017, USA

Adult Stem Cell Standardization / by Paolo Di Nardo.

Routledge is an imprint of the Taylor & Francis Group, an informa business

ISBN 978-87-92329-74-5 (print)

While every effort is made to provide dependable information, the publisher, authors, and editors cannot be held responsible for any errors or omissions.

Table of Contents

Preface

Stem cell research and technology represent a major challenge for treating otherwise non-curable patients. A decade of intensive research has demonstrated that initial hopes based more on enthusiasm than on solid scientific bases can be translated into factual techniques only by adopting more rigorous procedures and strategies. Among other major impediments, the failure so far experienced in applying stem cell technologies to repair parenchymal organs can be ascribed to the lack of sufficient knowledge on basic mechanisms, but also standardized criteria and protocols. Very often, each laboratory follows its own "recipe" using erratic nomenclature and non-comparable, if not confusing, experimental protocols. All this makes it difficult to learn from others and, ultimately, it hampers the progression of knowledge on stem cell behaviour.

The ambitious goal of this meeting was to gather the most innovative and scientifically robust knowledge and technologies on stem cells and to involve investigators from academy and industry in formulating recommendations to standardize the isolation and manipulation of stem cells using solid and well-documented knowledge rather than fragmentary and often unrepeatable experimental reports. This book collects some of the scientific contributions at the first Congress on Adult Stem Cell Standardization, held in Rome on December 1–3, 2010.

The Editor is very grateful to Valerio Alessandro Tavano and Fabio Spadaccini for their continued and invaluable organizational and editorial contribution.

List of Contributors

Albonici, L., *Division of Laboratory Medicine, Department of Experimental Medicine, University of Rome 2, Tor Vergata, Italy*

Barberi, L., *Institute Pasteur Cenci-Bolognetti, DAHFMO-Unit of Histology and Medical Embryology, IIM, University of Rome "La Sapienza", Rome, Italy*

Berta, G.N., *Department of Clinical and Biological Sciences, University of Turin, Turin, Italy*

Bontempo, L., *Division of Virology, Department of Experimental Pathology BMIE, University of Pisa, Pisa, Italy*

Caldarera, C.M., *Department of Biochemistry "G. Moruzzi" and National Institute for Cardiovascular Research (INRC), Università di Bologna, Bologna, Italy*

Cambi, C., *Unit of Dentistry and Oral Surgery, Department of Surgery, University of Pisa, Pisa, Italy*

Cassata, G., *Istituto Zooprofilattico Sperimentale della Sicilia, Palermo, Italy*

Ceccherini-Nelli, L., *Division of Virology, Department of Experimental Pathology BMIE, University of Pisa, Pisa, Italy*

Charbord, P., *INSERM U972, Hôpital de Bicêtre, University Paris 11, Le Kremlin Bicêtre, Paris, France*

De Luca, A., *Dipartimento BIONEC, Università degli Studi di Palermo, Italy*

De Rossi, M., *Institute Pasteur Cenci-Bolognetti, DAHFMO-Unit of Histology and Medical Embryology, IIM, University of Rome "La Sapienza", Rome, Italy*

Di Cola, G., *Department of Human Genetics Engineering, University of Parma, Parma, Italy; and EMI Group and European Clinic & Research ECR, Parma, Italy*

Di Felice, V., *Dipartimento BIONEC, Università degli Studi di Palermo, Italy*

Di Marco, P., *Istituto Zooprofilattico Sperimentale della Sicilia, Palermo, Italy*

Di Nardo, P., *Department of Internal Medicine, University of Rome 2, Tor Vergata, Rome, Italy*

Di Scipio, F., *Department of Clinical and Biological Sciences, University of Turin, Turin, Italy*

Dominguez-Bendala, J., *Diabetes Research Institute, University of Miami Millar School of Medicine, Miami, Florida, USA*

Folino, A., *Department of Clinical and Biological Sciences, University of Turin, Turin, Italy*

Fommei, E., *Gabriele Monasterio Foundation, Clinical Physiology Institute, National Research Council, CNR, Pisa, Italy*

Forte, G., *Department of Internal Medicine, University of Rome 2, Tor Vergata, Rome, Italy*

Gabriele, M., *Unit of Dentistry and Oral Surgery, Department of Surgery, University of Pisa, Pisa, Italy*

Gammazza, A.M., *Dipartimento BIONEC, Università degli Studi di Palermo, Italy*

Giordano, C., *Department of Biochemistry "G. Moruzzi" and National Institute for Cardiovascular Research (INRC), Università di Bologna, Bologna, Italy*

Govoni, M., *Department of Biochemistry "G. Moruzzi" and National Institute for Cardiovascular Research (INRC), Università di Bologna, Bologna, Italy*

Guarnieri, C., *Department of Biochemistry "G. Moruzzi" and National Institute for Cardiovascular Research (INRC), Università di Bologna, Bologna, Italy*

Guercio, A., *Istituto Zooprofilattico Sperimentale della Sicilia, Palermo, Italy*

Krishnan, L., *Thromosis Research Unit, Sree Chitra Tirunal Institute for Medical Sciences and Technology, Trivandrum, India*

Lepperdinger, G., *Extracellular Matrix Research Group, Institute for Biomedical Aging Research, Austrian Academy of Sciences, Innsbruck, Austria*

Liao, R., *Cardiac Muscle Research Laboratory, Cardiovascular Division, Department of Medicine, Brigham and Women's Hospital and Harvard Medical School, Boston, MA, USA*

Losano, G., *Department of Neurosciences, Physiology Division, University of Turin, Turin, Italy*

Magnani, E., *Laboratorio Cardiologia Molecolare & Cellulare, Dipartimento Medicina Interna, Università di Roma Tor Vergata, Rome, Italy*

Manzari, V., *Division of Laboratory Medicine, Department of Experimental Medicine, University of Rome 2, Tor Vergata, Italy*

Matteoli, B., *Division of Virology, Department of Experimental Pathology BMIE, University of Pisa, Pisa, Italy*

Minieri, M., *Laboratorio Cardiologia Molecolare & Cellulare, Dipartimento Medicina Interna, Università di Roma Tor Vergata, Rome, Italy*

Motta, A., *BIOtech Laboratories, Università degli Studi di Trento, Italy*

Musarò, A., *Institute Pasteur Cenci-Bolognetti, DAHFMO-Unit of Histology and Medical Embryology, IIM, University of Rome "La Sapienza", Rome, Italy; and Edith Cowan University, Western Australia*

Muscari, C., *Department of Biochemistry "G. Moruzzi" and National Institute for Cardiovascular Research (INRC), Università di Bologna, Bologna, Italy*

Nardone, G., *Laboratorio Cardiologia Molecolare & Cellulare, Dipartimento Medicina Interna, Università di Roma Tor Vergata, Rome, Italy*

Oikonomopoulos, A., *Cardiac Muscle Research Laboratory, Cardiovascular Division, Department of Medicine, Brigham and Women's Hospital and Harvard Medical School, Boston, MA, USA*

Pagliari, F., *Department of Clinical and Biological Sciences, University of Turin, Turin, Italy*

Pagliari, S., *Department of Internal Medicine, University of Rome 2, Tor Vergata, Rome, Italy*

Pagliaro, P., *Department of Clinical and Biological Sciences, University of Turin, Turin, Italy*

Parker, G., *The Carman and Ann Adams Department of Pediatrics, Wayne State University, School of Medicine, Children's Hospital of Michigan, Detroit, USA*

Pelosi, L., *Institute Pasteur Cenci-Bolognetti, DAHFMO-Unit of Histology and Medical Embryology, IIM, University of Rome "La Sapienza", Rome, Italy*

Pietronave, S., *Laboratorio Cardiologia Molecolare & Cellulare, Dipartimento Medicina Interna, Università di Roma Tor Vergata, Rome, Italy; and*

Department of Medical Sciences, Università del Piemonte Orientale "A. Avogadro", Novara, Italy

Prat, M., *Laboratorio Cardiologia Molecolare & Cellulare, Dipartimento Medicina Interna, Università di Roma Tor Vergata, Rome, Italy; and Department of Medical Sciences, Università del Piemonte Orientale "A. Avogadro", Novara, Italy*

Puleio, R., *Istituto Zooprofilattico Sperimentale della Sicilia, Palermo, Italy*

Rastaldo, R., *Department of Clinical and Biological Sciences, University of Turin, Turin, Italy*

Ricordi, C., *Diabetes Research Institute, University of Miami Millar School of Medicine, Miami, Florida, USA*

Rizzuto, L., *Dipartimento BIONEC, Università degli Studi di Palermo, Italy*

Roncoroni, L., *Surgical Clinic and Surgical Therapy, University of Parma, Parma, Italy*

Rosati, A., *Division of Virology, Department of Experimental Pathology BMIE, University of Pisa, Pisa, Italy*

Salamone, P., *Department of Clinical and Biological Sciences, University of Turin, Turin, Italy*

Sarli, L., *Surgical Clinic and Surgical Therapy, University of Parma, Parma, Italy*

Scaccino, A., *Division of Virology, Department of Experimental Pathology BMIE, University of Pisa, Pisa, Italy*

Schicchitano, B.M., *Institute Pasteur Cenci-Bolognetti, DAHFMO-Unit of Histology and Medical Embryology, IIM, University of Rome "La Sapienza", Rome, Italy*

Sereti, K.-I., *Cardiac Muscle Research Laboratory, Cardiovascular Division, Department of Medicine, Brigham and Women's Hospital and Harvard Medical School, Boston, MA, USA*

Serradifalco, C., *Dipartimento BIONEC, Università degli Studi di Palermo, Italy*

Sprio, A.E., *Department of Clinical and Biological Sciences, University of Turin, Turin, Italy*

Tobiasch, E., *University of Applied Sciences, Bonn-Rhein-Sieg, Germany*

Traversa, E., *Laboratorio Cardiologia Molecolare & Cellulare, Dipartimento Medicina Interna, Università di Roma Tor Vergata, Rome, Italy*

Verin, L., *BIOtech Laboratories, Università degli Studi di Trento, Italy*

Zamperone, A., *Department of Medical Sciences, Università del Piemonte Orientale "A. Avogadro", Novara, Italy*

Zhang, Y., *University of Applied Sciences, Bonn-Rhein-Sieg, Germany*

Zummo, G., *Dipartimento BIONEC, Università degli Studi di Palermo, Italy*

1

Adult Stem Cells Meet Three-Dimensional Culture Environments: A Perspective in Myocardial Tissue Restoring

Giovanni N. Berta[1], Raffaella Rastaldo[1], Federica Di Scipio[1],
Andrea E. Sprio[1], Paolina Salamone[1], Anna Folino[1],
Francesca Pagliari[2], Stefania Pagliari[2], Giancarlo Forte[2],
Pasquale Pagliaro[1], Paolo Di Nardo[2] and Gianni Losano[3]

[1]*Department of Clinical and Biological Sciences, University of Turin, Turin, Italy;
e-mail: giovanni.berta@unito.it*
[2]*Department of Internal Medicine, University of Rome 2, Tor Vergata, Rome, Italy*
[3]*Department of Neurosciences, Physiology Division, University of Turin, Turin,
Italy*

Abstract

Cell therapy is based on the concept that stem cells can proliferate and differentiate into specialized cells, to replace the dead ones and to allow the functional recovery of a damaged tissue. Although at the beginning this therapy appeared quite promising, real and long term results have not yet been achieved. The reason of many uncertain results is likely to be the lack of rigorous procedures clearly indicated by standardized protocols. This chapter is a contribution to the set up of a procedure for the selection of the cells and their implant in heart repair.

Keywords: adult stem cell, dental pulp mesenchymal stem cells, three-dimensional cell culture, myocardial tissue restoring.

P. Di Nardo (Ed.), Adult Stem Cell Standardization, 1–8.

1.1 Introduction

After a myocardial infarction, heart repair requires the use of stem cells with a good differentiation potential towards the various cardiac phenotypes, such as cardiomyocytes, endothelial and vascular smooth muscle cells (VSMC). Depending on their potentiality, stem cells may be *totipotent, pluripotent* and *multipotent.*

Totipotent stem cells are fertilized eggs and their first divisions. They can originate all embryonic tissues, from which all tissues and organs derive. Also extra-embryonic (e.g. placental) cells derive from totipotent cells.

Pluripotent stem cells are embryonic stem cells (ESC) which derive from the epiblast tissue of the inner cell mass (ICM) of a blastocyst or earlier morula stage embryos. They give rise to all derivatives of the three primary germ layers: ectoderm, endoderm and mesoderm. Some adult stem cells, present in the blood of the umbilical cord, may be classified as pluripotent [1].

With respect to the pluripotent cells, multipotent ones are lineage-restricted. They possess the ability to proliferate and create other cells like themselves (self-renewal) or, alternatively, to proliferate and give origin to more differentiated cells such as those of adipogenic, osteogenic, myogenic, endothelial, hepatic and also neuronal lineages [2, 3]. Fetal, amniotic and adult stem cells are included in this group. They may be found in bone marrow as mesenchymal stem cells (MSC) and hematopoietic stem cells, in adipose tissue as adipose-derived stem cells, in vascular endothelium as endothelial stem cells, in the heart as cardiac stem cells (CSC), etc. [4, 5].

1.2 Adult Stem Cells in the Cardiac Research Field

Due to the need of commitment towards the cardiac lineages, most of the studies on heart repair have been focused on adult stem cells, MSC and CSC. The latter have been classified in five groups, each of which is located in a specific cardiac region [6, 7]. It is likely that these five groups belong to the same population. In humans it has also been suggested that cardiomyocytes can reenter the cell cycle and proliferate [8].

Bone marrow MSC are likely to be attracted to the injured myocardium by stromal-cell derived factor 1 (SDF-1) released in the infarcted area [9]. It has also been reported that, after migration, MSC play mainly a trophic role by improving the performance of still viable tissue, while their proliferation and differentiation play a less important role in heart repair [9]. Obviously, in the case their migration is not sufficient to ameliorate the heart performance,

MSC can be taken from bone marrow after ischemia and reperfusion and implanted in the myocardium. Since they can be taken from the bone marrow of the same patient, the possibility of rejection is avoided.

Although CSC have the highest commitment towards cardiac phenotypes, they are not suitable for implantation because they should be taken from the same heart where they must be implanted, with an additional invasive intervention. As a consequence, the implantation is a procedure that mainly concerns MSC.

1.3 Mesenchymal Stem Cells from Dental Pulp

MSC may be isolated also from sources different from bone marrow, i.e. adipose tissue, umbilical cord and, more recently, dental tissues, such as dental pulp, apical papilla and periodontal ligament. They can also be obtained from human exfoliated deciduous teeth and dental follicle [10].

The main fate of dental pulp mesenchymal stem cells (DP-MSC) consists in their differentiation into odontoblasts, with a much more evident commitment than the corresponding cells of bone marrow. *In-vitro* DP-MSC behave like colony-forming unit-fibroblasts (CFU-F). These fibroblasts show dissimilar characteristics, while the different densities of the colonies suggest the possibility of different growth rates for various clones [11, 12]. In spite of this well defined commitment, they show also an excellent differentiation potential in response to specific *stimuli*.

Out of the odontoblast phenotype, *in-vitro* DP-MSC can differentiate towards adipose and neural cells, as well as osteogenic, condrogenic and myogenic lineages [10, 13–16]. It may be suggested that the differentiation versatility of these cells is related to the interaction of the epithelium of the first pharingeal arch with the neural crest as mesenchymal counterpart during the ontogenesis, which allows us to consider them as "ectomesenchymal" stem cells [10].

DP-MSC have been seen to differentiate into cardiomyocytes, thus suggesting the possibility of use in heart repair. In ongoing experiments we have seen that, independently of any treatment, these cells express both stemness and cardiomyocytes precursor markers. Since the main commitment of these cells is the continuous repair of dental tissues, one might argue that their differentiation concern dental rather than coronary vessels. However, the latter possibility cannot be excluded in the case of implantation in the heart wall.

So far the attempts to implant stem cells for heart repair have been mainly based on the injection near the infarcted area. The limited number of successes may be attributed to several factors, such as the trauma by the needle, the little, if any, confluence of the cells and the death which involves about 80% of the injected cells. Moreover, it is noteworthy that cells injected in a non-confluent or spread manner are responsible for an abundant release of vascular cell adhesion molecule-1 (VCAM-1) which mediates the adhesion of immune system cells to the endothelial ones, thus triggering a process leading to atherosclerosis.

An adequate confluence of the cells can be achieved with three-dimensional (3-D) "cell-sheets", as well as with 3-D scaffolds.

1.4 Three-Dimensional Cell Culture

It is important to make some preliminary remarks. In living animals all cells reside and interact with each other in a 3-D architecture of tissues and organs, which is critical for their growth and metabolism. Moreover, the phenotype and functions of individual cells are highly dependent on elaborate and sophisticated interactions with 3-D-organized extracellular matrix (ECM) proteins and neighboring cells [17]. Cell–ECM interactions are therefore of pivotal importance for normal cell differentiation and functions [18, 19]. *In-vitro* condition involves otherwise bi-dimensional (2-D) cell maintenance as monolayer, altering physiological cell-cell cross-talking, cell-ECM interactions and latterly the mechanical stimulations residing within the niche [20]. As a result, obtained data are often mere approximations, sometimes even unreliable. In addition, to allow tissue functional recovery, the use of monolayer cultured/differentiated stem cells is incompatible with the methods required to detach the cells from growing support and the intrinsic fragility of the cell monolayer itself. Although data accumulated over the past 30 years have demonstrated significant limitations in predicting the behavior of cells in living organisms, cell monolayer still remains the most popular model for *in-vitro* studies because of its easiness to handle, affordability and cheapness.

Furthermore, 2-D culture substrates not only prove inadequate to reproduce the complex and dynamic environments of *in-vivo* tissues, but they can also misrepresent the findings to some extent by forcing cells to adapt to an artificial, flat and rigid surface. Stiffness, in fact, is one of the most important factors to condition and maintain cell culture phenotype [21]. These cell–cell and cell–ECM interactions are considerably reduced or even absent in the case of 2-D cell culture on a flat substrate, which in turn significantly limits their ability to recapitulate the appropriate level of *in-vivo* cellular responses.

Thanks to the rapid progress in tissue engineering and new micro-scale technologies, conventional problems based on 2-D cell monolayers can be fruitfully overcome by innovative 3-D cell-based models, which are more adequate to create a physiologically appropriate, reproducible and well controlled cell-based system for more reliable experiments free of the problems mentioned above [22]. In fact, in the last decade a wide spectrum of 3-D matrices (surfaces/scaffolds), characterized by specific stiffness, composition and geometry, have been designed to mimic the particular physiological microenvironment with the aim to reduce the gap between cell cultures and living tissues. "Skin equivalents" represent the most diffuse and successful 3-D organotypic models that, effectively, have been productively used in pharmaco-toxicological studies or for grafting procedures [23]. Unfortunately, the ability to produce bioengineered tissues is inversely proportional to their complexity. An example is just the myocardium, characterized by unique biological characteristics and mechanical properties.

Moreover, the attempts to obtain bioengineered cardiac tissue were mostly frustrated by the inability to maintain in culture mature and differentiated myocytes for adequate periods. In fact, the only line that can be used must be not yet mature and not susceptible to premature senescence [24].

Many research groups reported the possibility to cultivate on scaffolds various cell types (neonatal, fetal or embryonic cardiomyocytes, and various embryonic and adult stem cell populations) [25], while no attempt has so far been made with cardiac stem/progenitor cells. Several clinical trials using stem/progenitor cells are in progress, but the interaction of such cells with the scaffolds still represents an intriguing but far option. Furthermore, the pre-clinical validation of 3-D biocompatible, biodegradable scaffolds with defined size and shape, controlled porosity and myocardial-like stiffness as tools to pre-commit human stem cells has never been performed so far. Therefore, the use of stem/progenitor cells may be able to overcome all these problems. An interesting aspect may be the possibility to prepare personalized pre-committed, cardiac bio-substitutes using autologous cells: this could revolutionize future clinical treatments of cardiac diseases by allowing to set up cardiac patches *ex-vivo* using autologous stem cells. Indeed these cells can be characterized for the correct genetic and antigenic expressions throughout the process of extraction, expansion and differentiation, with a fully standardized procedure.

1.5 Conclusion

In conclusion, these types of engineered myocardial bio-substitutes may be produced by means of autologous stem cells and biocompatible and bio-degradable scaffolds, manufactured to match myocardial properties and to reduce host immune reaction thus improving cardiac function. Indeed, a standardized proposed approach is expected to have an impact on overall healthcare costs (chronic treatments, (multiple) surgical treatments, period of hospitalization, informal care, etc.) and to decrease the number of patients waiting for heart transplantation because of the reduction of heart allograft requirements. Standardized procedures in stem cell culture should allow to start all the studies with a well characterized and classified population and to compare the results in response to different protocols. In the absence of an appropriate standardization, it remains difficult, if not impossible, to compare the results of different investigations as regards cell selection, implant and clinical follow-up.

References

[1] Ratajczak, M.Z., Machalinski, B., Wojakowski, W., Ratajczak, J., Kucia, M. A hypothesis for an embryonic origin of pluripotent Oct-4(+) stem cells in adult bone marrow and other tissues. *Leukemia*, 21(5):860–867, May 2007.

[2] De Coppi, P., Barstch, G., Atala, A. Isolation of amniotic stem cell lines with potential for therapy. *Nature Biotechnology*, 25(5):100–106, May 2007.

[3] Jiang, Y., Jahagirdar, B.N., Reinhardt, R.L., Schwartz, R.E., Keene, C.D., Ortiz-Gonzalez, X.R., Reyes, M., Lenvik, T., Lund, T., Blackstad, M., Du, J., Aldrich, S., Lisberg, A., Low, W.C., Largaespada, D.A., Verfaillie, C.M. Pluripotency of mesenchymal stem cells derived from adult marrow. *Nature*, 418(6893):41–49, July 2002.

[4] Barrilleaux, B., Phinney, D.G., Prockop, D.J., O'Connor, K.C. Review: Ex vivo engineering of living tissues with adult stem cells, *Tissue Engineering*, 12(11):3007–3019, November 2006.

[5] Gimble, J.M., Katz, A.J., Bunnell, B.A. Adipose-derived stem cells for regenerative medicine, *Circulation Research*, 100(9):1249–1260, May 2007.

[6] Di Nardo, P., Forte, G., Ahluwalia, A., Minieri, M. Cardiac progenitor cell: Potency and control, *Journal of Cell Physiology*, 224(3):590–600, September 2010.

[7] Loewy Kirby, M. *Cardiac Development*. Oxford University Press, Oxford, 2007.

[8] Kajstura, J., Leri, A., Finato, N., Di Loreto, C., Beltrami, C.A., Anversa, P. Myocyte proliferation in end-stage cardiac failure in humans, *Proceedings of the National Academy of Sciences USA*, 95(15):8801–8805, July 1998.

[9] Zhang, M., Mal, N., Kiedrowski, M., Chacko, M., Askari, A.T., Popovic, Z.B., Koc, O.N., Penn, M.S. SDF-1 expression by mesenchymal stem cells results in trophic support of cardiac myocytes after myocardial infarction. *FASEB Journal*, 21(12):3197–3207, October 2007.

[10] Huang, G.T., Gronthos, S., Shi, S. Mesenchymal stem cells derived from dental tissues vs. those from other sources: Their biology and role in regenerative medicine. *Journal of Dental Research*, 88(9):792–806, September 2009.

[11] Gronthos, S., Mankani, M., Brahim, J., Robey, P.G., Shi, S. Postnatal human dental pulp stem cells (DPSCs) in vitro and in vivo, *Proceedings of National Academy of Sciences USA*, 97(25):13625–13630, December 2000.

[12] Huang, G.T., Shagramanova, K., Chan, S.W. Formation of odontoblast-like cells from cultured human dental pulp cells on dentin in vitro. *Journal of Endodontic*, 32(11):1066–1073, November 2006.

[13] Gronthos, S., Brahim, J., Li, W., Fisher, L.W., Cherman, N., Boyde, A., DenBesten, P., Robey, P.G., Shi, S. Stem cell properties of human dental pulp stem cells. *Journal of Dental Research*, 81(8):531–535, August 2002.

[14] Laino, G., D'Aquino, R., Graziano, A., Lanza, V., Carinci, F., Naro, F., Pirozzi, G., Papaccio, G. A new population of human adult dental pulp stem cells: A useful source of living autologous fibrous bone tissue (LAB). *Journal of Bone Mineral Research*, 20(8):1394–1402, August 2005.

[15] Zhang, W., Walboomers, X.F., Shi, S., Fan, M., Jansen, J.A. Multilineage differentiation potential of stem cells derived from human dental pulp after cryopreservation. *Tissue Engineering*, 12(10):2813–2823, October 2006.

[16] D'Aquino, R., Graziano, A., Sampaolesi, M., Laino, G., Pirozzi, G., De Rosa, A., Papaccio, G. Human postnatal dental pulp cells co-differentiate into osteoblasts and endotheliocytes: A pivotal synergy leading to adult bone tissue formation, *Cell Death Differentiation*, 14(6):1162–1171, June 2007.

[17] Lund, A.W., Yener, B., Stegemann, J.P., Plopper, G.E. The natural and engineered 3D microenvironment as a regulatory cue during stem cell fate determination. *Tissue Engineering Part B Review*, 15(3):371–380, September 2009.

[18] Justice, B.A., Badr, N.A., Felder, R.A. 3D cell culture opens new dimensions in cell-based assays. *Drug Discovery Today*, 14(1/2):102–107, January 2009.

[19] Lee, J., Cuddihy, M.J., Kotov, N.A. Three-dimensional cell culture matrices: State of the art. *Tissue Engineering Part B Review*, 14(1):61–86, March 2008.

[20] Mazzoleni, G., Di Lorenzo, D., Steimberg, N. Modelling tissues in 3D: The next future of pharmaco-toxicology and food research. *Genes and Nutrition*, 4(1):13–22, March 2009.

[21] Yeung, T., Georges, P.C., Flanagan, L.A., Marg, B., Ortiz, M., Funaki, M., Zahir, N., Ming, W., Weaver, V., Janmey, P.A. Effects of substrate stiffness on cell morphology, cytoskeletal structure, and adhesion. *Cell Motility and the Cytoskeleton*, 60(1):24–34, January 2005.

[22] Maltman, D.J., Przyborski, S.A. Developments in three-dimensional cell culture technology aimed at improving the accuracy of in vitro analyses. *Biochemical Society Transaction*, 38(4):1072–1075, August 2010.

[23] Brohem, C.A., Da Silva Cardeal, L.B., Tiago, M., Soengas, M.S., De Moraes Barros, S.B., Maria-Engler, S.S. Artificial skin in perspective: Concepts and applications. *Pigment Cell Melanoma Research*, 24(1):35–50, February 2011.

[24] Leor, J., Amsalem, Y., Cohen, S. Cells, scaffolds, and molecules for myocardial tissue engineering. *Pharmacological Therapy*, 105(2):151–163, February 2005.

[25] Baek, H.S., Park, Y.H., Seok, K.C., Park, J.C., Rah, D.K. Observation of proliferation and attachment of three different cell types on nano-fibrous silk film & scaffold. *Key Engineering Materials*, 342/343:85–88, July 2007.

2

Should Viral Un-infectibility Be an Additional Required Characteristic of a Deliverable Adult Progenitor Cell?

Barbara Matteoli[1], Loredana Albonici[2], Giancarlo Forte[3], Lucia Bontempo[1], Antonio Scaccino[1], Antonella Rosati[1], Christian Cambi[4], Mario Gabriele[4], Enza Fommei[5], Vittorio Manzari[2], Paolo Di Nardo[2] and Luca Ceccherini-Nelli[1]

[1]*Division of Virology, Department of Experimenal Pathology BMIE, University of Pisa, Pisa, Italy; e-mail: luca.ceccherini-nelli@med.unipi.it*
[2]*Division of Laboratory Medicine, Department of Experimental Medicine, University of Rome 2, Tor Vergata, Rome, Italy*
[3]*Laboratory of Molecular and Cellular Cardiology, Department of Internal Medicine, University of Rome 2, Tor Vergata, Rome, Italy*
[4]*Unit of Dentistry and Oral Surgery, Department of Surgery, University of Pisa, Pisa, Italy*
[5]*Gabriele Monasterio Foundation, Clinical Physiology Institute, National Research Council, CNR, Pisa, Italy*

Abstract

Stem cells can represent a privileged target for noxious agents, such as viruses, which, possibly reactivated in the host by immune suppression, could invalidate implanted engineered tissues vanishing the cell therapy advantages. In search for biomarkers within the cellular specific expression pathways that define the cell "non-infectibility" status, we reviewed known stem cell susceptibility to viral infection, host genetics and epigenetic reprogramming of host genes in viral infection and we have indicated molecular pathways now known to be involved by viral infections (in particular, the mechanisms

P. Di Nardo (Ed.), Adult Stem Cell Standardization, 9–30.

presiding over the generation of the viral innate immunity, such as viral PAMPs, TLRs, apoptotic signaling pathways, inflammasome biology, interferon induction and negative regulation) whose knowledge we believe will be critical to the selection of stem cell populations with natural or induced resistence to viral infection to be used for replacement therapy.

Keywords: virus, virus infection, infectibility, un-infectibility, adult progenitor cell, replacement therapy, regenerative medicine, host genetics, epigenetics, viral PAMPs, TLRs, apoptotic signaling pathways, inflammosome biology, interferone induction pathways.

2.1 Introduction

Much attention is currently paid to cell determinants that identify the optimal stem cell population and to factors and mechanisms inducing their appropriate differentiation. However, stem cells could also represent a privileged target for noxious agents, such as viruses, which, possibly reactivated by immune suppression, could invalidate implanted engineered tissues vanishing the cell therapy advantages. Therefore, knowledge must be accumulated in order to define stem cell susceptibility to viral infections. This information will be crucial to develop novel procedures to isolate stem cell populations resistant to viral infections (numerous viruses are capable of persisting within specific tissues) and to manage the follow-up of patients treated with cell therapy in the long term.

Under experimental *in vivo* and *in vitro* conditions of infection, stem cells display low susceptibility to viral replication that becomes more evident as differentiation increases [21]. Stem cell viral competence, however, once thought to be strictly correlated to receptor expression, is now believed to be also dissociated from it [2]. It is, therefore, necessary to identify new biomarkers within the cellular specific expression pathways that define the cell "non-infectibility" status.

In this chapter, the stem cell sources, main applications and the host genetics and epigenetic reprogramming of host genes in viral infection are reviewed to define the feasibility of novel procedures allowing the isolation of stem cell populations with favorable characteristics. The reported viral persistence [16] in hosts, tissues and stem cells will also be analyzed to determine whether the viral infection can be reactivated by immune suppression and can cause infection of the replacing stem cells. Finally, possible molecular pathways involved by viral infections will be proposed, since their understanding could be critical to select stem cell populations with natural or induced resistence to viral infection to be used for replacement therapy.

2.2 Stem Cell Sources and Applications

2.2.1 Human Embryonic Stem Cells (hESCs)

They have the greatest potential for regenerative medicine, but, unfortunately, they have limited application because of ethical and technical problems, including rejection, if the patient is not immunologically compatible, and the consequent need for immune-suppressing treatments [4].

2.2.2 Viral Induced Pluripotent Stem (Viral iPS) Cells

Obtained by viral transduction with four transcription factors, OCT-4, SOX2, C-MYC and KLF4, of the patient's somatic cells. They have unlimited self-renewal and the potential to differentiate into all somatic cell types. Their greatest benefit is that, since they are derived from the patient, there are no immunological compatibility problems, allowing a patient-specific therapy. At present, they represent the best candidates for a replacement therapy [20].

2.2.3 CD34+ Cord Blood Cells Obtained by Inhibition of p53

The introduction of TP53 shRNA drastically increases the number of ES cell-like clones that can be derived from CD34+ cells, but the beneficial effect of p53 repression on the reprogramming efficiency may reflect the survival of cells containing genetic lesions that would be otherwise eliminated through apoptosis. It is not clear yet whether this can be considered a safe method to increase the efficiency of stem cell pluripotency, but, since cord blood cells are obtained shortly after birth, they are less likely to contain potentially dangerous mutations [22].

2.2.4 mTERT-Sca-1pos-bone Marrow Murine/Human Stromal Cell Obtained by Inhibition of Telomerase

Mesenchymal stem cells, MSC, have been transfected with pCINeo-mTERT to obtain Telomerase inhibition (dubbed mTERT-MSC) by transfecting purified Sca-1pos mesenchymal progenitors isolated from the murine bone marrow with the catalytic subunit of the telomerase enzyme [6], a procedure that has already been proven effective in extending lifespan in embryonic and adult stem cells. The mTERT-MSC line behaves exactly like native, freshly isolated, MSC.

2.2.5 Adult Stem Cells

Clusters of stem cells are resident virtually in all organs. In order to stimulate their activation for therapeutic uses, growth factors have been administered *in vivo* into the damaged organs. Alternatively, isolated resident stem cells have been injected into the damaged organs or have been seeded *in vitro* on biodegradable, biocompatible scaffolds that are meant to provide physical support to cell adhesion, proliferation, and differentiation, while favoring the organization of newly-formed tissues. However, scaffold properties can exert different effects on distinct stem cell populations; in fact, bone marrow-derived mesenchymal stem cells (MSC) do not activate the same gene programs as cardiac stem cells (CSC), when cultured in the same conditions (scaffold/culture medium), thus suggesting that MSC and CSC do not share all their functional features, the latter being more prone to acquire a specific (i.e. the cardiac) phenotype, when challenged by an adequate microenvironment.

Several fields of regenerative medicine are involved in early and advanced experimental situations that prelude a clinical use. For instance *human severe combined immunodeficiencies* (SCIDs) or other *primary immunodeficiencies* (non-SCID PIDs), where hematopoietic stem cell transplantation for primary immunodeficiency before 6 months of age is associated with improved outcome and supports the use of newborn screening programs to facilitate the early diagnosis of SCID [7].

Adult T-cell leukemia (ATL), a lymphoproliferative malignancy associated with human T-cell lymphotropic virus type 1 (HTLV-1) infection, recently, has been investigated for a curative treatment with *allogeneic hematopoietic stem cell transplantation* (allo-HSCT). HTLV-1 Tax-specific CD8+ cytotoxic T cells (CTL) responses against viral Tax have also been shown to play a critical role in the control of HTLV-1 virus replication *in vivo*, and CTLs contribute to the graft-versus-ATL effect [23].

Stem cells from a donor who was homozygous for a 32-bp deletion in the CCR5 allele, known to provide genetic resistance to HIV-1 infection, were transplanted in a patient with *acute myeloid leukemia and HIV-1 infection*: the patient remained without viral rebound 20 months after transplantation and discontinuation of antiretroviral therapy [10]. This outcome demonstrates the critical role that CCR5 plays in maintaining HIV-1 infection, and moreover this could be a potential avenue for controlling HIV without antiretroviral therapy, leading possibly to a prospect for an HIV cure.

In a *viral-induced demyelization mouse model of Multiple Sclerosis*, mouse neural stem cells (NSCs) or human oligodendrocyte progenitor cells (OPCs) derived from human embryonic stem cell (hESC) were used to promote remyelination in mice persistently infected with the neuroadapted JHM strain of mouse hepatitis virus (MHV), undergoing tissue damage because of immune-mediated demyelination [24]; here stem cell transplantation results in migration, proliferation, and differentiation of the cells into OPCs and mature oligodendrocytes that is associated with increased axonal remyelination possibly through the CXCL12:CXCR4 pathway capable of regulating homing of engrafted stem cells to sites [1].

In *experimental Coxsackievirus B3 (CVB3)-induced myocarditis*, initially considered a sole immune-mediated disease, now known to result also from a direct CVB3-mediated injury of the cardiomyocytes, mesenchymal stem cells (MSCs) have been proven to have, besides immunomodulatory, also anti-apoptotic features. MSCs could be infected by CVB3, could reduce the direct CVB3-mediated cardiomyocyte injury and viral progeny release, *in vitro*, in the absence of immune cells; it has been suggested that *in vivo* MSC application could improve murine acute CVB3-induced myocarditis [26].

2.3 Host Genetics and Epigenetic Reprogramming of Host Genes in Viral Infection

Viruses are obligate intracellular pathogens, by necessity highly pre-adapted to developing ways of hijacking cell processes to facilitate the completion of their life cycle and survive the host immune response. Since virus–host interaction is an important determinant of disease outcome, the net result of this interaction is an adaptive host–viral relationship that is crucial for shaping both host and pathogen genetic diversity: analyses of susceptibility or severity of infection, time to clinical outcome (i.e. time to AIDS or death following HIV infection; outcome of infection, treatment response and major complications such as hepatocellular carcinoma or fibrosis in persistent hepatitis) have been interrelated to multiple phenotypes.

Evolutionary relationships between host genetic factors and viral infections are bidirectional: viruses influence human genetic diversity just as human genetic factors influence the outcome of viral infections. Host genetic variation is appreciated now beyond the single genotype level, incorporating extended haplotypes [13] as well as regions of segmental genetic duplication: an increasing number of polymorphic immune factors, including main MHC alleles, killer immunoglobulin-like receptors and functional chemokine receptor polymorphisms have been reported [12].

2.3.1 HIV

The role of variation in the CCR5 gene in resistance to infection and slower disease progression is now well established [10]. Another HIV-1 co-receptor ligand in HIV infection, stromal derived factor 1 (SDF1), is also associated with HIV-1 disease progression in some populations. Delayed progression to AIDS has been linked to the flanking CCR2 gene, in a single amino acid change, while increased disease progression to AIDS was associated in both European and African Americans, with the CCR5 ligand RANTES, and the presence of an intronic SNP that differentially binds regulatory proteins suggesting an evolutionarily important role for this new SNP in immunomodulation. Many other genes have been identified as HIV-1/AIDS host genetic determinants. These include genes such as IL4, IL10 and NRAMP1 that have been linked also with other infectious diseases.

2.3.2 HBV, HCV

Infection with either the hepatitis B or hepatitis C virus results in either an acute, self-limited disease or, in a minority, in persistent infection. Persistent carriage rates, which confer an increased risk of liver complications, failure or end stage carcinoma, are 10–20% in hepatitis B when compared with 80–90% of hepatitis C infections. Studies of the host genetic factors implicated in disease chronicity have found that non-HLA genes such as TNFA, MBL and VDR, are all associated with persistence of hepatitis B infection. The unusual immunological dichotomy of the outcome of disease makes HBV viral persistence an ideal candidate for family linkage-based studies enabling the identification of novel major genes that determine this fate. A genomewide scan in Gambian families has mapped a major susceptibility locus to chromosome 21 and two neighboring genes appear to be involved.

In HCV viral persistence, the high numbers that are treated (with either interferon alpha alone or in combination with antiretroviral therapy) allow for the analysis of genetic influence on treatment response and complications of infection in addition to the outcome of infection. The role of the CCR5 receptor, already acknowledged to be important in HIV viral infection, has been investigated with respect to HCV infection. Initially involved in HCV persistence, it seems now more likely that CCR5 may play a role in the risk of liver complications such as portal inflammation and risk of fibrosis but not to the outcome of infection. Studies of the interferon induced genes such as MxA and PKR show that variation in these genes is a factor in the hosts' response to therapy. Even though persistence rates are higher in HCV than in HBV, the lower prevalence makes family identification difficult, therefore limiting the possibility of understanding in family studies the genetics of HCV persistence.

The majority of infectious diseases' host susceptibility is likely to be highly polygenic. Disease pathogenesis, that has been evaluated to date only with a few dozen candidate genes, may be now analyzed by the International HapMap project method [9, 13], by scanning an entire human genome sequence, or by increasing amounts of resequencing data from chromosomal segments, or by directly scanning millions of SNPs available for association studies.

Mammalian gene regulation can be involved in stable, heritable, covalent modifications initiated and established with the repression or silencing of gene transcription (epigenetic: a stably heritable phenotype resulting from changes in a chromosome without alterations in the DNA sequence, through alterate methylation, acetylation, phosphorylation and ubiquitylation patterns on histones and DNA) that can be involved in the development of human diseases such as cancer or host responses associated with immunity and inflammation or chronic disease. To establish and mediate epigenetic memory, such modifications must be transmitted during DNA replication. Some viruses and bacteria are known or thought to induce epigenetic changes in host cells possibly in chronic diseases associated with microbial persistence. Two human γ-herpesviruses, Kaposi's sarcoma associated virus (KSHV or HHV8) and the Epstein–Barr virus (EBV) are the best examples to illustrate how epigenetic manipulation of host cells probably contributes to latency and is likely to be involved in disease pathogenesis. Hepatitis B virus (HBV), Human Papillomavirus (HPV), Adenoviruses, HIV and Human T cell leukaemia virus-1 (HTLV-1), Simian vacuolating virus 40 (a Poliomavirus that infects monkeys and humans), are less well-characterized examples. Host genes involved in cell cycle progression, senescence, survival, inflammation and immunity are prime candidates as targets for such epigenetic control [14].

2.4 Viral Persistence, Stem Cells Infectibility

Viruses attach to host cells by binding to receptors on the cell surface; entry occurs via receptor-mediated viral translocation, direct or receptor-mediated endocytosis, or through membrane fusion after a sufficient number of receptors have engaged ligand proteins on the virion. Under conditions where the cell surface receptor densities are low, recruitment of receptors may be limited by diffusion rather than by receptor-ligand interactions [8].

2.4.1 HBV, HCV

More than 500 million people worldwide are persistently infected with either hepatitis B virus (HBV) or hepatitis C virus (HCV). Although both viruses are poorly cytopathic, persistent infection causes severe immunopathologic damage to liver tissue; histologically, such damage is characterized by fatty liver disease, liver fibrosis, and a higher likelihood of hepatocellular carcinoma. Virus-specific CD8+ T cells play a crucial role during infection with hepatitis viruses. On one hand, rapid activation of CD8+ T cells can control the virus and therefore inhibit its persistence. On the other hand once the virus persists in the liver, the chronic activation of virus-specific T cells leads to continued liver cell damage. This double-edged role of CD8+ T cells determines the final outcome of infection. In half of the cases of human HCV infection, the virus persists; in the other half, the virus is controlled.

2.4.2 Parvovirus

B19 human parvovirus is the etiologic agent of erythema infectiosum, transient aplastic crisis, virus-induced hematopoietic suppression, a polyarthralgia syndrome in adults, and some cases of hydrops fetalis. Hematopoietic progenitor cells can be serially assayed in B19-infected and uninfected bone marrow cultures. At initiation, B19 virus infection causes marked and moderate reduction in colony-forming unit erythroid (CFU-E) and burst-forming unit erythroid (BFU-E) numbers, respectively, without affecting CFU-Mix and CFU-GM numbers. Interestingly, the recovery of the erythroid progenitor numbers can be observed at a late stage.

However, in experimental B19 infection of volunteers, erythroid progenitors in the bone marrow does not always decrease in number despite the disappearance of erythroblasts at the time of viremia. Similarly, in a case of chronic bone marrow failure due to persistent B19 virus infection, the number of erythroid progenitors in the bone marrow is normal. This discrepancy between the *in vivo* and *in vitro* data has remained unexplained. Moreover, in situ hybridization of bone marrow cells infected with B19 virus show that 30 to 40% of the erythroblasts contained the viral genome, suggesting a rather broad host cell range within the erythroid cell lineage. The target cells of B19 virus are in the erythroid lineage from BFU-E to erythroblasts, with susceptibility to the virus increasing along with differentiation. Furthermore, the suppression of erythropoiesis and the subsequent recovery of the erythroid progenitor numbers in B19-infected liquid cultures may be analogous in part to the clinical features of B19 virus-induced transient aplastic crisis [21].

2.4.3 JC Polyoma Virus

Human embryonic stem cell-derived oligodendrocyte progenitor cells are susceptible to infection with the JC virus, the causative agent of progressive multifocal leukoencephalopathy (PML). A human embryonic stem cell line, H7, can be used to derive an enriched population of cells expressing the oligodendrocyte progenitor cell-specific marker NG2. These cells express the 5HT2a receptor (5HT2aR) for JC virus and are highly susceptible to infection. Infection can be reduced by treatment with anti-5HT2aR antibodies and by the 5HT2aR antagonists ritanserin and ketanserin. This demonstrates that human embryonic stem cell-derived oligodendrocyte progenitor cells are susceptible to JC virus infection and indicates that cells poised to replenish mature oligodendrocytes in PML lesions may also be a target of viral infection [19].

2.4.4 Junin Virus

The Human Immunodeficiency Virus (HIV), human herpesvirus 6 and human cytomegalovirus negatively affect the survival, differentiation and/or maturation of megakaryocyte progenitors derived from CD34+ cells: in contrast, JUNV infection had no significant effect on the proliferation, clonogenic ability or maturation of megakaryocyte progenitors, but it induces a profound decrease in proplatelet formation and platelet release. TfR1 is the receptor for some new world arenaviruses, including JUNV: blocking or down-regulating TfR1 surface expression in CD34+ cells significantly reduces both JUNV replication as well as the decrease in platelet formation induced by JUNV infection, indicating that viral infection is a necessary event for the inhibition of platelet formation. Moreover, the fact that the UV-irradiated virus has no effect on proplatelet production and platelet release indicates that viral replication is also necessary to hinder thrombopoiesis. These data also implicate TfR1 as the main route of viral entry into hematopoietic progenitor cells. This is particularly relevant considering that it has recently been shown that pathogenic strains can use TfR1-dependent or independent pathways. The observation that JUNV infection up-regulates TfR1 in CD34+ cells could represent a viral dissemination strategy at least in hematopoietic bone marrow cells. The analysis of the signaling pathways involved in the JUNV-mediated inhibition of platelet production indicates that SOCS-1, SOCS-3, GATA-1, and Src, which are involved in the processes of proliferation, differentiation, and proplatelet formation, respectively, are not involved. In contrast, JUNV-infected samples show lower levels of both NF-E2 mRNA and protein. Interestingly, type I IFN downregulates NF-E2 in megakaryocytes. As this transcription factor plays a major role in terminal differentiation of megakaryocytes and platelet release, it is conceivable that the reduced synthesis of NF-E2 mediated by type I IFN may be one of the molecular pathogenic mechanisms reducing platelet release in JUNV infection. In fact, thrombocytopenia driven by recombinant IFN-α has been related to the selective inhibition of cytoplasmic maturation accompanied by downregulation of the expression of the transcription factors NF-E2, GATA-1 and MafG/HPRT [15].

2.4.5 CMV

Congenital cytomegalovirus (CMV) infection is a significant cause of brain disorders, such as microcephaly, mental retardation, hearing loss and visual disorders in humans. The type and severity of brain disorder may be dependent on the stage of embryonic development when the congenital infection occurs. Developmental disorders may be associated with the type of embryonic cells to which CMV is susceptible and the effects of the infection on the cellular functions of these cells. Early murine embryos, including embryonic stem (ES) cells, are not susceptible to CMV infection. A part of the embryonic cells acquire susceptibility during early development. Mesenchymal cells are the targets of infection at midgestation, affecting organogenesis of the brain, eyes and oral-facial regions. In contrast to ES cells, neural stem progenitor cells (NSPC) from fetal brains are susceptible to murine CMV (MCMV) infection. The viral infection inhibits proliferation and differentiation of the NSPC to neuronal and glial cells in addition to induction of neuronal cell loss. These cellular events may cause brain malformations, such as microcephaly and polymicrogyria. Furthermore, MCMV persists in neuronal cells in developing brains, presumably resulting in neuronal dysfunction [25].

2.4.6 Coxsackievirus B3

RD cells are often considered to be non-permissive for infection with coxsackievirus and adenovirus receptor (CAR)-dependent group B coxsack-ieviruses (CVB): inoculated cell monolayers show little or no cytopathic effect (CPE) and immunohistological assays for CAR have been consistently negative. Supernatants recovered from RD cells exposed to CVB, however, containes more virus than was added in the initial inoculum, indicating that productive virus replication occurres in the monolayer. When infected with a recombinant CVB type 3 (CVB3) chimeric strain expressing S-Tag within the viral polyprotein, 4–11% of RD cells expressed S-Tag over 48 h. CAR mRNA was detected in RD cells by RT-PCR, and CAR protein was detected on Western blots of RD lysates; both were detected at much lower levels than in HeLa cells. Receptor blockade by an anti-CAR antibody confirmed that CVB3 infection of RD cells was mediated by CAR. These results show that some RD cells in the culture population express CAR and can thereby be infected by CVB, which explains the replication of CAR-dependent CVB in cell types that show little or no CPE and in which CAR has not previously been detected. Cells within cultures of cell types that have been considered non-permissive may express receptor transiently, leading to persistent replication of virus within the cultured population [2].

WAO9, HUES-5 and HUES-16 human embryonic stem cell lines and derived contractile embryoid bodies have been investigated for susceptibility to Coxsackievirus B infection. After validating stem cell-like properties and cardiac phenotype, Coxsackievirus B receptors CAR and DAF, as well as type I interferon receptors were detected in all cell lines and differentiation stages studied. Real-time PCR analysis showed that CAR mRNA levels were 3.4-fold higher in undifferentiated cells, while DAF transcript levels were 2.78-fold more abundant in differentiated cultures ($P < 0.05$). All cell lines were susceptible to Coxsackievirus serotypes B1–5 infection as shown by RT-PCR detection of viral RNA, immunofluorescence detection of viral protein and infectivity titration of cell culture supernatants resulting in cell death. Supernatants infectivity titers 24–48 h post-infection ranged from 10^5–10^6 plaque forming units (PFU/ml), the highest titers were detected in undifferentiated cells. Cell viability detected by a colorimetric assay showed inverse correlation with infectivity titers of cell culture supernatants.

Treatment with 100 U of interferon Iβ significantly reduced viral replication and associated cell death during a 24–48 h observation period, as detected by reduced infectivity titers in the supernatants and increased cell viability by a colorimetric assay, respectively [18].

Enteroviral infection of the heart has been noted in a significant proportion of cases of myocarditis and dilated cardiomyopathy. The presence of enterovirus RNA at stages of disease after acute infection and correlation of enterovirus replication with worse clinical outcome suggests that continued replication of the virus is involved in the progression of the disease. This finding is mirrored by the murine model of coxsackievirus B3 myocarditis, in which virus persists through the evolution of the virus to a terminally deleted defective form which persists in the myocardium. Studies of the mechanism of induction of myocarditis by coxsackievirus B3 require assessment of the effects of alterations of the immune response upon virus persistence in this form. As expression of viral proteins in the heart have been shown to generate significant impairment of cardiomyocyte function and promote generation of dilated cardiomyopathy, the role of virus persistence is likely to include direct effects of viral replication as well as induction of inflammation in the heart. Factors that control the extent of cardiac infection with terminally deleted enteroviruses and the relative roles of continued immune response of the virus versus viral modification of cardiac function need to be measured to find effective therapies for the human disease.

2.5 Which Potential Markers of Un-infectibility Could Be Possibly Used in the Selection of the Perfect Deliverable Adult Progenitor Cell?

Increasing evidence points to the activation of functional intracellular pathways that correlate with virus ability of host infection and clinical end result of that viral infection. Cell-virus interactions are investigated within the intracellular sensors of innate immunity: (1) *Viral Pathogen-Associated Molecular Patterns* (PAMPs) that initiate signaling cascades that lead to the production of pro-inflammatory cytokines, (2) *Toll-like receptors* (TLRs) and related activated cytokine responses, (3) *Death-Receptor mediated Apoptotic signaling pathways*, (4) *Inflammosomes and NLRs virus immune evasion*, and (5) *Interferon-regulatory factor 3*.

The selection of stem cells that after differentiation may still show naturally impairment of such mechanisms may lead to the identification of stem cells resistant to viral infection; alternatively intervention in stem cells on those molecular pathways may block the process of host infection after replacement therapy.

2.5.1 Viral Pathogen-Associated Molecular Patterns (PAMPs)

Viral PAMPs are composed mainly of unique nucleic acids, such as double-stranded (ds)RNA, uncapped single-stranded (ss)RNA and cytosolic DNA, but also include several viral fusion glycoproteins (for example, respiratory syncytial virus (RSV)-F protein and Ebola virus GP1 protein).

Four main classes of PRRs have been described: Toll-like receptors (TLRs), retinoic acid inducible gene-I (RIG-I)-like receptors (RLRs), nucleotide binding oligomerization domain (NOD)-like receptors (NLRs) and the IFI200 family member absent in melanoma 2 (AIM2). Whereas TLRs sense PAMPs in the extracellular space and endosomes, NLRs, RLRs and AIM2 function as pathogen sensors in intracellular compartments. TLR interact with products of infectious agents to activate cells of the innate immune system and also stimulate the adaptive immune system. Intracellular and extracellular TLRs recognise a wide range of viruses leading to the production of different cytokines. Innate immune activation by viruses is triggered by transmembrane TLRs both on the cell surface (e.g. TLR2 and TLR4) and within endosomal/ER intracellular compartments (e.g. TLR9, TLR7/8 and TLR3). In the cytosol, helicases such as RIG-I function as sensors for viral replication intermediates. TLRs and the cytosolic helicases activate type I IFN and inflammatory cytokine production via IRF and NF-kB. The TLRs have a cytosolic signalling motif, the TIR domain, which is essential for activation of shared downstream adapters (MyD88, Mal/TIRAP and TRIF) and activation of intracellular signal transducing proteins including IRAK/IKK$\alpha\beta\gamma$ and TBK-1/IKKε/IRF3 signalling cascades.

Viral pathogen-associated molecular patterns (PAMPs) activate nucleotide-binding oligomerization domain (NOD)-like receptors (NLRs) and inflammasomes to initiate signaling cascades that lead to the production of pro-inflammatory cytokines, thereby amplifying antiviral innate immune responses. The end result is the phosphorylation and activation of interferon (IFN) response factor 3 (IRF3) and IRF7 to turn on the transcription of type I IFN (IFNα/β) genes. The NLRs NOD2, NLR family member X1 (NLRX1) and NLR family CARD-containing protein 5 (NLRC5) associate with MAVS. Whereas NOD2 mediates the induction of type I IFNs, NLRX1 and NLRC5 inhibit RIG I-MAVS interactions and thereby negatively regulate type I IFN production [11].

2.5.2 Toll-Like Receptors (TLRs) and Related Activated Cytokine Responses

TLRs are "pattern recognition proteins" that discriminate between self and non-self.

The outcome of virus-TLR interactions depends on both the type of virus and its expression of TLR ligands (i.e. virion/capsid proteins, nucleic acid replication intermediates, etc.) and on the pattern of TLR expression on host cells. Herpes Simplex Virus (HSV) activates two distinct TLRs, TLR2 on the cell surface and TLR9 within the ER. TLR2 is expressed by monocyte/macrophage populations and triggering of this receptor by HSV virions results in a NF-kB dependent inflammatory cytokine secretion response. TLR9 is expressed by plasmacytoid dendritic cells and HSV activates these cells via its CpG-rich DNA genome resulting in high levels of IFN-α and cytokine secretion. TLR9 (as well as MyD88) is also involved in the development of CD8+ CTL responses. Viruses with their intracellular lifecycle are capable of stimulating through intracellular as well as extracellular TLRs. One type of virus would be necessarily more likely than another to interact with a given TLR [5]. The complexity of viral proteins makes it likely that many will be found to interact with many different TLRs (although not all TLRs have been proven, as yet, to interact with viral proteins). The ability of reoviruses (dsRNA viruses) to stimulate TLR3, which is also activated artificially by the double stranded RNA poly I:C, seems a very predictable result; it seems less obvious that the West Nile virus (a single stranded RNA virus) has a clear TLR3-mediated effect, and that the pathogenesis of the murine cytomegalovirus (MCMV), a DNA virus, is influenced by both TLR3 and TLR7. However, based on the fact that DNA viruses and positive strand RNA viruses, like dsRNA viruses, have a dsRNA intermediary, it is predictable that many of these viruses will have TLR3-specific activating activity, but this has not been proven for all such viruses. At the same time it would appear that negative strand viruses (like influenza and the vesicular stomatitis virus, VSV) which are capable of activating TLR7, may not be able to interact with TLR3 because these viruses do not have a double stranded RNA phase as part of their lifecycle. As TLR9 interacts with DNA sequences containing CpG motifs, it is not surprising that several DNA viruses have been shown to stimulate cytokine production through a TLR9 dependent pathway. HSV will induce interferon through TLR9 which is expressed on pDC. Interestingly, however, HSV will also stimulate TLR2 expressed on the surface of macrophages. Since TLR2 is expressed in brain cells as well, induction of cytokines through TLR2 may lead to encephalitis [5].

In addition to stimulating TLR signalling events, both through their proteins and nucleic acids, viruses have been demonstrated to inhibit TLR signalling. Vaccinia virus protein A46R, and its homologous A52R protein expression inhibited TLR-mediated activation of NF-κB, and activation mediated by other TLR ligands; mutant virus isolates, lacking these proteins, are less virulent, underscoring the major role these proteins have in viral pathogenesis. HCV expresses a virally encoded serine protease (NS3/4A) that has been demonstrated to specifically degrade TRIF. This results in inhibition of TLR3 mediated-signalling which is thought to be important in the ability of this virus to maintain itself as a chronic infection. Interestingly, at the same time that the virus inhibits TLR4-mediated events, it has been demonstrated to sensitize certain cells for enhanced production of cytokines by increasing expression of TLR4 so illustrating the complexity of virus-TLR interactions.

Activation of TLRs is critical for virus specific long-term immunity so the lack of a TLR response may lead to defects in the virus specific immune response. TLR9 activating compounds could increase virus specific immunity and, in work as vaccine adjuvants. TLR3 agonists have been demonstrated to reduce genital herpes in animal models when applied topically. However, TLR activation could act as a "double-edged sword" and possibly turn off the "cytokine storm" that accompanies hemorrhagic fever viruses and is responsible for the damage associated with viral encephalitis.

2.5.3 Death-Receptor Mediated Apoptotic Signalling Pathways

Programmed cell death may occur by activation of death receptor (extrinsic)- or mitochondrial (intrinsic)-mediated signaling pathways. Extrinsic apoptotic signaling involves the activation of cell surface death receptors belonging to the tumor necrosis factor (TNF) receptor (TNFR) family of proteins, including Fas/APO-1, TNFR-1, and TNF-related apoptosis inducing ligand receptors 1 and 2 (TRAIL-R1 and TRAIL-R2). These receptors are activated after binding of their cognate ligands namely, Fas ligand (FasL), TNF, and TRAIL. Death receptors contain a cytoplasmic death domain (DD) that serves as a docking site for homotypic DD interactions with DD-containing adaptor proteins. Fas-associating protein with a death domain (FADD) is the adaptor protein for Fas and is recruited to the activated receptor along with procaspase 8 to form a death-induced signaling complex. Caspase 8 is activated at the death-induced signaling complex and can then activate downstream effector caspases (such as caspase 3), resulting in apoptosis. The crucial role of FADD and caspase 8 in Fas signaling is shown in FADD- or caspase 8-deficient mice that are resistant to Fas-induced apoptosis. The intrinsic apoptotic pathway involves the release of pro-apoptotic factors through pores in the mitochondrial membrane. Pro-apoptotic factors released through mitochondrial pores include cytochrome *c*, which triggers the activation of caspase 9 and second mitochondrion-derived activator of caspases (SMAC), which downregulate cellular inhibitor of apoptosis proteins (IAPs). Mitochondrial pores consist of dimers of pro-apoptotic members of the Bcl-2 family of proteins (Bax and Bak) and are tightly regulated by interactions with other (anti-apoptotic and BH3-only) Bcl-2 family proteins. Apoptosis has been identified as an important mechanism of cardiac myocyte death in experimental models of viral myocarditis, as well as in endomyocardial biopsies from patients with viral myocarditis. The death receptor associated initiator caspase, caspase 8, and the effector caspase, caspase 3, were reported significantly activated after infection of primary cardiac myocytes with myocarditic, but not non-myocarditic, reovirus strains. Furthermore, reovirus-induced cardiac myocyte apoptosis was significantly inhibited by soluble death receptors. In contrast, the mitochondrial membrane potential remained unaltered and caspase 9, the initiator caspase associated with mitochondrial apoptotic signaling, was only weakly activated in cardiac myocytes after infection with myocarditic reovirus strains. Inhibition of mitochondrial apoptotic signaling had no effect on reovirus-induced cardiac myocyte apoptosis. Therefore both *in vivo* and *in vitro* data, would infer that caspase 8, but not caspase 9, results significantly activated in the hearts of reovirus-infected mice [3]. In addition, reovirus-induced injury can be effectively ameliorated by inhibition of apoptosis-associated cysteine protease activation (calpain and caspase) both *in vitro* and *in vivo*.

2.5.4 Inflammosomes and NLRs Virus Immune Evasion

Host proteins that are involved in the recognition of virus infection have only recently been identified and the initial characterization of their roles in this complex system is only just beginning to be understood. Identification of the NLR and RLR family of pathogen sensors in the past several years has provided important information concerning the cellular recognition of viruses. NLRs and inflammasomes are necessary for the innate inflammatory control of virus infection, as well as for healing responses. In addition, the fact that multiple viruses encode inflammasome inhibitors is consistent with the importance of this pathway for inhibiting virus replication.

In the evolutionary race between a pathogen and its host, the host has undergone selection for various mechanisms to detect the presence of viruses and closely regulates the production of master cytokines such as IL1β and IL18 by the inflammasome. In response, viruses have undergone selective pressure to evade activation of the inflammasome. Several viruses are known to encode proteins that interfere with inflammasome signalling. Inflammasome signalling is disrupted at the level of the adaptor protein ASC or caspase1 itself, as both of these molecules are activated downstream of multiple NLRs, as well as downstream of AIM2 and RIG1 inflammasomes. The large genomes of poxviruses encode multiple inhibitors that interfere with innate and adaptive immunity. Bioinformatics studies show that many poxviruses encode a putative PYD-containing protein that is hypothesized to interact with ASC; however, the only examples that have been studied in detail are the myxoma virus M13L-PYD and Shope Fibroma virus S013L proteins. M13L-PYD is required for the pathogenesis of myxoma virus and deletion of this protein results in severe virus attenuation *in vivo*. This is shown by a decrease in viraemia owing to inefficient replication in lymphocytes and leukocytes and increased inflammation at the initial site of infection, which together lead to more rapid resolution of disease. *In vitro*, infection of cells with myxoma virus that lacks M13L-PYD leads to increased activation of caspase-1 and increased levels of IL-1β and IL-18; these findings indicate that M13L-PYD has an immunosuppressive function. M13L-PYD and S013L associate with ASC through direct PYD-PYD interactions.

Expression of either M13L-PYD or S013L alone inhibits caspase-1 activation and IL-1β production downstream of the NLRP3 inflammasome in cell culture. Poxvirus PYD-containing proteins can therefore inhibit inflammasome activation at the level of the adaptor protein ASC and prevent PRRs from activating caspase-1 in virus-infected cells. In addition, ASC might have several caspase-1-independent immune functions, such as the activation of T cells and lymph node cells; it is therefore possible that poxvirus PYD-containing proteins block inflammation by inhibiting these processes as well.

Poxviruses also encode various serpin-like protease inhibitors (SPIs) that inhibit caspase-1. For example, CrmA (also known as SPI-2) of the cowpox virus is a wellknown caspase-1 inhibitor. CrmA deficiency decreases the number and severity of pox lesions on the allantoic membrane when the cowpox virus is grown in chicken eggs. Respiratory tract infection of mice with cowpox virus or rabbit poxvirus mutants that lack CrmA results in decreased inflammation and viral loads compared with wild-type virus. After intradermal inoculation with cowpox virus, more rapid virus clearance and a more robust inflammatory response are observed with virus lacking CrmA. It is interesting to note that no differences in IL-1β levels were observed in the lungs of C57BL/6 mice that were infected with wildtype cowpox virus versus mutant virus lacking CrmA, even though *in vitro* studies indicate that CrmA can inhibit IL-1β production and secretion.

Other poxviruses also encode SPI homologues, such as Serp2 in myxoma virus. Deletion of Serp2 results in severe attenuation of myxoma virus infection in rabbits. Some vaccinia viruses also encode an SPI-2 protein. However, deletion of vaccinia virus SPI-2 had no effect on IL-1β-induced fever responses in infected mice and vaccinia virus SPI-2 mutants were not attenuated *in vivo*. Instead, fever reduction and attenuated weight loss in vaccinia virus-infected mice depended on vaccinia virus-encoded IL-1β scavenger receptor (vIL-1βR).

Several poxviruses also encode an IL-18-binding protein that blocks IL-18-induced signalling. It is becoming clear that poxviruses have developed various inflammasome inhibitors and molecules that interfere with downstream signalling induced by the inflammasome products IL-1β and IL-18.

However, viral inflammasome inhibitors are not limited to poxviruses. The NS1 protein of influenza A/PR/8/34 H1N1 virus prevents caspase-1 activation and IL-1β production. The N-terminus of this NS1 protein was also suggested to interfere with inflammasome activation, although the details remain unclear. Influenza A/PR/8/34 H1N1 virus that lacks the N-terminus of NS1 is attenuated in cell culture and induces higher levels of IL-1β and cell death. Interestingly, inhibition of the dsRNA-dependent protein kinase PKR also decreases IL-1β production in cells that are infected with NS1-mutant influenza A/PR/8/34 H1N1 virus. Although these results do not provide a clear inhibitory mechanism for NS1, they do show the importance of this protein for virus replication. However, the ability of NS1 to block caspase-1 activation seems to be strain specific, as NS1 from highly pathogenic H5N1 bird influenza A virus activates caspases and induces apoptosis. One possible explanation for the ability of H5N1 influenza A virus to grow despite activating caspases is that it downregulates NLRP3 and IL-1β expression levels by 24 hours after infection and therefore H5N1 virus might not require inhibition of caspase-1 to suppress inflammation. By contrast, the influenza A/PR/8/34 H1N1 virus has been shown to upregulate the expression of inflammasome components and might therefore require NS1-mediated inhibition of caspase-1 to prevent inflammation and to replicate successfully. It is probable that there are other viral immune evasion pathways that target inflammasome components.

Therapeutic interventions that are designed to target the virus–host interface of such immune evasion strategies to restore the functions of PRRs and their signaling pathways will offer the potential to restore innate immune induction and inflammatory programmes that initiate immunity for the control of virus infection (for a review, see [11]).

2.5.5 Interferon-Regulatory Factor 3

Increasing evidence has shown the importance of pattern-recognition receptors in immune responses after viral and microbial infection by invading pathogens. Toll-like receptor 3 (TLR3) detects extracellular viral double-stranded RNA (dsRNA) internalized into the endosomes, whereas retinoic acid-inducible gene I (RIG-I), a DExD/H box RNA helicase containing a caspase recruitment domain, detects intracellular viral dsRNA. TLR4, in contrast, recognizes microbial components such as bacterial lipopolysaccharide (LPS). Engagement of any of those receptors triggers rapid production of type I interferon (IFN-$\alpha\beta$) and thus establishes the innate immune status against infectious agents. Interferon-regulatory factor 3 (IRF3), a ubiquitously expressed transcription factor, is responsible for the primary induction of IFN-β and is important in the establishment of innate immunity in response to either viral or microbial infection. After the detection of pathogens, IRF3 is phosphorylated on multiple phosphorylation acceptor (phospho-acceptor) sites, forms homodimers and then translocates to the nucleus, where it binds to the interferon stimulation-response elements of target genes, as well as the positive regulatory domain III-I in the IFN-b promoter. The mechanisms underlying the phosphorylation-induced activation of IRF3 have been the subject of many extensive studies. The substitution of alanine for either the Ser385 or Ser386 residue of IRF3 abolishes its activation. Additionally, phosphorylation of Ser386 on IRF3 is induced by TLR3 engagement and by viral infection and only for IRF3 dimers. The importance of five critical serine or threonine residues of IRF3 (Ser396, Ser398, Ser402, Thr404 and Ser405) for its activation has been demonstrated. Notably, the substitution of alanine for all five amino acids abrogates the function of IRF3 to activate transcription, whereas aspartic acid substitutions result in a constitutively active protein. Those published data demonstrate that phosphorylation of both C-terminal phospho-accepter clusters (Ser385-Ser386 and Ser396-Ser398-Ser402-Thr404-Ser405) is important for the activation of IRF3. Other studies have also shown that two IkB kinase (IKK)-like kinases, TBK1-NAK and IKK-i–IKKe, are required for the activation of IRF3 by inducing the phosphorylation of its two C-terminal phospho-acceptor clusters and thus are essential in the expression of type I interferon. Phosphorylation-dependent post-translational modifications of IRF3 are therefore crucial for regulating the function of IRF3.

Pin1 is a peptidyl-prolyl isomerase that via its WW domain (with two conserved tryptophan residues) specifically recognizes phosphorylated serine or threonine residues followed by proline and then catalyzes a conformational change of the bound substrate in a phosphorylation-dependent way. By that mechanism, Pin1 has been shown to regulate the stability and or localization of its substrates during transcriptional activation, cell cycle progression and cell death, and deregulated expression or loss of function of Pin1 leads to the progression of important human diseases such as cancer and Alzheimer disease. However, a regulatory function for Pin1 in the host defense against infectious agents and associated signal transduction pathways has not been reported before to our knowledge. Published findings showing that post-translational modification of IRF3 by phosphorylation controls the IRF3 activity prompted us to assess the involvement of Pin1 in regulating IRF3 signaling [17].

2.6 Concluding Remarks

The application of cell treatments to regenerate injured organs holds great promises, but is a field that is still in its infancy. Many questions about the potential effects induced in stem cells by inherent and environmental factors still need to be addressed. Among many others, the relationship virus-stem cell remains totally unknown, but of fundamental relevance to control adverse reactions to cell implants. In particular, the mechanisms presiding over the generation of the viral innate immunity, such as viral PAMPs, TLRs, apoptotic signaling pathways, inflammasome biology, interferon induction and negative regulation must be in-depth investigated, before, within the host variability, stem cells populations displaying resistance to infectibility can be safely isolated and exploited for clinical uses.

References

[1] Carbajal K., Schaumburg C., Strieter C., Kane J., and Lane T.E. Migration of engrafted neural stem cells is mediated by CXCL12 signaling through CXCR4 in a viral model of multiple sclerosis. *PNAS*, 107:11068–11073, 2010.

[2] Carson S.D., Kim K.S., Pirruccello S.J., Tracy S., Chapman N.M. Endogenous low-level expression of the coxsackievirus and adenovirus receptor enables coxsackievirus B3 infection of RD cells. *J. Gen. Virol.*, 88:3031–3038, 2007.

[3] Debiasi R., Robinson B.A., Leser S., Brown D., Long C., Clarke P. Critical role for death-receptor mediated apoptotic signaling in viral myocarditis. *J. Cardiac. Fail.*, 16:901–910, 2010.

[4] Ensenat-Waser R., Pellicer A., Simon C. Reprogrammed induced pluripotent stem cells: How suitable could they be in reproductive medicine? *Fertility and Sterility*, 91:971–974, 2009.

[5] Finberg R.W., Wang J.P., Kurt-Jones E.A. Toll like receptors and viruses. *Rev. Med. Virol.*, 17:35–43, 2007.

[6] Forte G., Franzese O., Pagliari S., Pagliari F., Di Francesco A.M., Cossa P., Laudisi A., Fiaccavento R., Minieri M., Bonmassar E., Di Nardo P. Interfacing Sca-1(pos) mesenchymal stem cells with biocompatible scaffolds with different chemical composition and geometry. *J. Biomed. Biotechnol.*, 2009:1–10, 2009.

[7] Gennery A.R., Slatter M.A., Grandin L., Taupin, P., Cant, A.J., Veys, P. et al. Transplantation of hematopoietic stem cells and long-term survival for primary immunodeficiencies in Europe: Entering a new century, do we do better? *J. Allergy Clin. Immunol.*, 126:602–611, 2010.

[8] Gibbons M.M., Chou T., D'Orsogna M.R. Diffusion-dependent mechanisms of receptor engagement and viral entry. *J. Phys. Chem. B*, 114(46):15403–15412, 2010.

[9] Sun P., Zhang R., Jiang Y., Wang X., Li J., Lv H., Tang G., Guo X., Meng X., Zhang H., Zhang R. Assessing the patterns of linkage disequilibrium in genic regions of the human genome. *FEBS J.*, August 9. doi: 10.1111/j.1742-4658.2011.08293.x. [Epub ahead of print], PMID: 21824289, 2011.

[10] Hütter G., Nowak D., Mossner M., Ganepola S., Müssig A., Allers K., Schneider T., Hofmann J., Kücherer C., Blau O., Blau I.W., Hofmann W.K., Thiel E. Long-term control of HIV by CCR5 Delta32/Delta32 stem-cell transplantation. *New Engl. J. Med.*, 360:692–698, 2009.

[11] Kanneganti T.D. Central roles of NLRs and inflammasomes in viral infection. *Nature*, 10:688–698, 2010.

[12] Nolan D., Gaudieri S., Mallal S. Host genetics and viral infections: Immunology taught by viruses, virology taught by the immune system. *Curr. Opin. Immunol.*, 18:413–421, 2006.

[13] Ong R.T.H., Liu X., Poh W.T., Sim X., Chia K.S., Teo Y.Y. A method for identifying haplotypes carrying the causative allele in positive natural selection and genome-wide association studies. *Bioinformatics*, 27:822–828, 2011.

[14] Paschos K., Allday M.J. Epigenetic reprogramming of host genes in viral and microbial pathogenesis. *Trends Microbiol.*, 18:439–447, 2010, review.

[15] Pozner R.G., Ure A.E., De Giusti C.J., D'Atri L.P., Italiano J.E., Torres O., Romanowski V., Schattner M., Gómez R.M. Junín virus infection of human hematopoietic progenitors

impairs *in vitro* proplatelet formation and platelet release via a bystander effect involving type I IFN signaling. *PLoS Pathogens*, 6:1–14, 2010, e1000847.

[16] Richard S. Hypothesis: Replicative homeostasis. A fundamental mechanism mediating selective viral replication and escape mutation. *Virology J.*, 2:10–14, 2005.

[17] Saitoh T., Tun-Kyi A., Ryo A., Yamamoto M., Finn G., Fujita T., Akira S., Yamamoto N., Lu KP., Yamaoka S. Negative regulation of interferon-regulatory factor 3-dependent innate antiviral response by the prolyl isomerase Pin1. *Nature Immunology*, 7:598–605, 2006.

[18] Scassa M.E., de Giusti C.J., Questa M., Pretre G., Richardson G.A., Bluguermann C., Romorini L., Ferrer M.F., Sevlever G.E., Miriuka S.G., Gómez R.M. Human embryonic stem cells and derived contractile embryoid bodies are susceptible to Coxsakievirus B infection and respond to interferon Iβ treatment. *Stem Cell Res.*, 6:13–22, 2011.

[19] Schaumburg C., O'Hara B.A., Lane T.A., Atwood W.J. Human embryonic stem cell-derived oligodendrocyte progenitor cells express the serotonin receptor and are susceptible to JC virus infection. *J. Virology*, 82:8896–8899, 2008.

[20] Takahashi K., Yamanaka S. Induction of pluripotent stem cells from mouse embryonic and adult fibroblast cultures by defined factors. *Cell*, 126:663–676, 2006.

[21] Takahashi T., Ozawa K., Takahashi K., Asano S., Takaku F. Susceptibility of human erythropoietic cells to B19 parvovirus *in vitro* increases with differentiation. *Blood*, 75:603–610, 1990.

[22] Takenaka C., Nishishita N., Takada N., Jakt L.M., Kawamata S. Effective generation of iPS cells from CD34+ cord blood cells by inhibition of p53. *Exp. Hematol.*, 38:154–162, 2010.

[23] Tanaka Y., Nakasone H., Yamazaki R., Sato K., Sato M., Terasako K., Kimura S., Okuda S., Kako K., Oshima K., Tanihara A., Nishida J., Yoshikawa T., Nakatsura T., Sugiyama H., Kanda Y. Single-cell analysis of T-cell receptor repertoire of HTLV-1 tax-specific cytotoxic T cells in allogeneic transplant recipients with adult T-Cell leukemia/lymphoma. *Cancer Res.*, 70:6181–6192, 2010.

[24] Tirotta E., Carbajal K.S., Schaumburg C.S., Whitman L., Lane T.E. Cell replacement therapies to promote remyelination in a viral model of demyelination. *J. Neuroimmunol.*, 224:101–107, 2010.

[25] Tsutsui Y. Effects of cytomegalovirus infection on embryogenesis and brain development. *Congenital Anomalies*, 49:47–55, 2009.

[26] Van Linthout S., Savvatis K., Miteva K., Peng J., Ringe J., Warstat K., Schmidt-Lucke C., Sittinger M., Schultheiss H.P., Tscho C. Mesenchymal stem cells improve murine acute coxsackievirus B3-induced myocarditis. *European Heart Journal*, Advance Access, published 22 December 2010.

3

Recent Results on Human Bone Marrow Mesenchymal Stem Cell Biology

Pierre Charbord

INSERM U972, Hôpital de Bicêtre, University Paris 11, Le Kremlin Bicêtre, Paris, France; e-mail: pcharbord@noos.fr

Abstract

This chapter recapitulates the major biological attributes of human bone marrow Mesenchymal Stem Cells: self-renewal, multipotency and lineage priming, and hematopoietic support. It also indicates the properties of the bone marrow native cells that are amplified according to specific culture conditions. Standardization of the culture procedure is essential for future experimental and clinical studies using these cells.

Keywords: stem cell, differentiation, bone marrow, cell culture, vasculature.

3.1 Introduction

In this chapter I will review some of the recently solved issues on the biology of Bone Marrow Mesenchymal Stem Cells. These issues concern some of the major questions raised by this very specific type of cells, i.e.:

- Standardization of the culture procedure.
- Self-renewal.
- Differentiation potential.
- Membrane phenotype.
- Nature and isolation of the native cells.

P. Di Nardo (Ed.), Adult Stem Cell Standardization, 31–40.

3.2 Previous Data and Controversies

Bone Marrow Mesenchymal Stem Cells (BM MSCs) are the precursors of the skeletal connective tissue-forming cells, including the three mesenchymal lineages (Osteoblastic, Chondrocytic and Adipocytic) and the family of Vascular smooth muscle cells (mural cells of arterioles, capillaries and marrow sinuses).

BM MSCs have been first described in the 1950s by Alexander Friedenstein who isolated, from the bone marrow of a number of animals (mouse, rat, guinea pig), cells forming colonies of fibroblasts, consequently named CFU-Fs. CFU-Fs were able to give rise to bone and to fibrous tissue both in *in vitro* and *in vivo* assays. Moreover, some of the donor CFU-Fs developed under the renal capsule were described as the homing sites for recipient circulating Hematopoietic Stem Cells (HSC) and were therefore able to harbor hematopoietic tissue generated from these. Later studies have shown that CFU-Fs gave rise also to chondrocytes and adipocytes, identifying therefore CFU-Fs as BM MSCs. For further reading, see [1].

The different cell types generated by BM MSCs constitute the non-hematopoietic compartment of bones, and BM MSCs form the stromal component of the Hematopoietic Stem Cell niche. Indeed two distinct niches have been described [2–4]. In the endosteal niche, the HSC is directly in contact with the bony surface or located within the first cellular layers from this surface. In the vascular niche, the HSC is close to the endothelium lining capillaries or bone marrow sinuses. Stromal cells in the niche (either endosteal or vascular) insure the proper functioning of the HSC, i.e. the adequate balance between self-renewal and commitment to the hematopoietic lineages.

The four-fold differentiation potential of MSCs is well established. A myriad of studies have shown that culture-amplified cells can, upon shifting of the culture conditions, differentiate into A, O and C. A more limited number of reports has shown that MSCs can also differentiate *in vitro* into V cells [5–10]. The in vitro data have been complemented by *in vivo* studies. Many studies have shown that human BM MSCs ectopically implanted under the skin of immune-deficient mice were able to give rise to bone and to adipocytes. Likewise these cells seeded in a collagen-fibronectin matrix implanted into a cranial window of immune-deficient mice contributed, in the presence of endothelial cells, to the formation of stable vascular networks [5].

A number of studies have suggested that BM MSCs may have a larger differentiation potential including additional mesodermal (endothelium and skeletal and cardiac sarcomeric muscle), neuro-ectodermal (neurons and astrocytes) and endodermal (hepatocytic) lineages (for a review see [1]). Such large differentiation potential would render BM MSCs akin to pluripotent stem cells.

Whether BM MSCs constitute a *bona fide* stem cell population has been the matter of debate. The crux of the debate resided in the demonstration of the self-renewal capacity. In the absence of clearcut evidence of this paradigmatic property, the International Society for Cellular Therapy has proposed in 2005 to call these cells Mesenchymal Stromal Cells, which maintained the acronym MSC while modifying its meaning, shifting the emphasis from stemness to hematopoietic support [11]. It is somewhat confusing that for many investigators "stromal" refers to the well known immune-modulatory properties of the cells instead of their hematopoietic supportive capacity.

Many of the controversies concerning BM MSC properties may be related to their culture amplification, which raises two issues, that of the standardization of the culture procedures and that of the isolation of native cells that could be used as substitute of cultured cells in future experimental procedures and clinical trials.

3.3 Recent Results

3.3.1 Establishment of a Standardized Culture Procedure

Within the European consortium "Genostem" including more than 30 European teams, we strictly defined protocols for *in vitro* amplification and differentiation of BM MSCs [9, 12, 13].

Bone marrow was collected usually from iliac crest and cells were seeded in alpha-MEM medium without nucleosides, but supplemented with 10% Fetal Calf Serum and 1 ng/mL basic fibroblast growth factor (FGF2). The FCS was screened to provide optimal growth, but not to enhance differentiation in any lineages. The sample did not undergo any treatment before seeding: no density separation and no red cell lysis. The flasks were not coated with extra-cellular matrix molecules. The cell density was 50,000 viable cells/cm^2 at culture inception and 1000 cells/cm^2 at each passage. The non-adherent cell fraction was discarded by day 3 by totally replacing the supernatant by fresh medium. The cultures were incubated in normoxic condition. Cells were passaged when layers reached confluence, after 2–3 weeks in culture. One to two passage(s) yielded sufficient number of viable mesenchymal cells (devoid of hematopoietic contaminant) for future studies. Cells were then frozen at 10^6 cells per vial in 10% DMSO.

Clones were generated at culture inception by seeding 100 mono-nucleated cells/cm^2. After 3–5 days, single clones were isolated by cloning rings and seeded in 35 mm plates. In the next days, cells were passaged into flasks of increasing surface, the timing of passages being monitored from the rapidity of growth. It took generally for well-growing clones about 21 days to reach a cell number similar to that obtained from polyclonal primary layers. Well-growing clones constituted no more than a third of the total number of clones, indicating that most clones grew poorly and were therefore unavailable for future studies. Clones generated at passage 1 or 2 were generally of small size, yielding a few hundred cells. Although numerable as CFU-Fs, they were too small to be used for subsequent studies.

For differentiation, cells were passaged in differentiation media. The critical inducers in these media were dexamethasone, dexamethasone + isobutyl-methyl-xanthine + indomethacin, high concentration of Transforming Growth Factor-β (e.g. 10 ng/mL of TGFβ1), and cytokines provided by the long-term culture medium, for O, A, C and V, respectively. Another critical factor for C was hypoxia and tight cell contact achieved through aggregate formation. Differentiation was obtained after 14, 21, 21 and 28 days for A, O, C and V, respectively.

This set of conditions was applied throughout the entire consortium so that similar results were obtained from one group to the other.

3.3.2 Self-Renewal Capacity

The team of Paolo Bianco, a partner team in Genostem, has evidenced self-renewal capacity [14]. CD146+/CD90+ cells from one clone were admixed with hydroxyapatite-calcium phosphate scaffolds and implanted subcutaneously into immune-deficient mouse. A few weeks later the mouse was euthanatized, the implant was recovered, and the minute fraction of CD146+/CD90+ cells was sorted by FACS and seeded to new culture plate. After a few days two clones were generated, indicating that the initial colony-forming cell had divided into two clonogenic cells.

This demonstration is essential for the qualification of BM MSCs as stem cells. The self-renewal capacity has been recently demonstrated also for mouse BM MSCs [15]. These data conclude the long-standing debate on MSC stemness.

3.3.3 Differentiation Potential

We have studied, in BM MSC clones, transcripts coding for 30–60 markers of 10 different lineages, not only A, O, C and V but also endothelial, sarcomeric (skeletal and cardiac) muscle, hepatocytic, neural and pluripotency markers (300 transcripts in all) [9]. Results have shown that, before any differentiation induction, many A, O, C and V markers were expressed, while only half the endothelial markers were detected and most markers of other lineages were undetectable. Of particular relevance was the detection of the master transcription factors RUNX2, SOX9 and PPARG critical for the commitment of MSCs into O, C and A lineages, respectively. These data indicate that MSCs are primed to the mesenchymal and vascular smooth muscle lineages, i.e. express at low level critical transcripts characteristic of the lineages to which they can commit. On the contrary, the quasi-absence of transcripts of other lineages (except for endothelial cells) indicates that differentiation into these pathways is not a physiological attribute of these cells, i.e. cannot be induced by the usual signaling pathways triggered by the binding of morphogens or cytokines to their cognate receptors. Differentiation to the non-primed lineages should therefore be enforced by non-physiological means, such as the use of agents that de-methylate methyl-cytosines in CpGs islands of promoters (e.g. 5-azacytidine), or gene transfer of critical transcription factors. Such procedures induce partial or total reprograming of the genetic pathways, similarly to what is done when generating induced pluripotent cells [16].

Our data emphazise another similarity between BM MSCs and other types of stem/progenitor cells, since lineage priming is a recognized feature of HSCs and embryonic stem cells [17].

Our data are not in contradiction with previous reports indicating larger differentiation potential, even confining to pluripotency. Several factors have to be taken into consideration. Foremost are culture conditions, which vary to a large extent from an investigator to the other. Some differences are evident such as the use of specific culture media or batch of sera, low or high seeding density at culture inception or at passages, initial selection of cells (e.g. by their resistance to stress). Other differences may be subtle but significant, such careful monitoring of the time of passaging before full confluence had been reached. Finally, biological parameters should be taken into account such the body origin of cells. It has indeed been shown in the mouse that the first wave of MSCs detected at early developmental stage was of neuro-ectodermic instead of mesodermic origin [18]; moreover, it has been reported that a fraction of adult mouse BM MSCs were still neuro-ectodermic in origin [19]. Such origins may explain the neural differentiation of BM MSCs reported by certain investigators. It is expected that MSCs taken from the upper part of the body (like dental sheath) may differ in their differentiation potential from MSCs from lower part (like iliac crest). Moreover, it is known that mesenchymal cells retain some imprinting from their site of development [20], which may also impact on their differentiation ability.

These considerations imply that any qualification of cells as MSCs must not only rely on the description of their phenotypic and functional properties, but must also include the precise protocol used to obtain and amplify them. Hence our endeavor to standardize the culture conditions before any attempt to define the generated cells.

3.3.4 Membrane Phenotype

Although it is usually believed that the BM MSC phenotype is neither hematopoietic nor endothelial, no extensive studies had been made on their membrane phenotype. We have made such a genome-wide study at the transcript level, and confirmed at the protein level, by flow cytometry, the presence or absence of many molecules [21]. Among transcripts coding for outer plasma membrane proteins, we have detected the expression of 464 molecules, i.e. 28.5% of all outer plasma membrane proteins inventoried in the microarray. As expected 22% of the detected molecules consisted of transcripts coding for cell adhesion molecules (mainly integrins, tetraspanins, surface proteoglycans and molecules of the immunoglobulin superfamily), while 15% consisted of transcripts coding for morphogen and cytokine receptors (primarily wnt, notch, bone morphogenetic protein and tyrosine kinase receptors). As expected also, very few transcripts for interleukin receptors, receptors for hormones or neuro-mediators, and receptors expressed on immune cells, were detected. Due to antibodies availability, we could study 115 proteins, 51 of which were detected at significant level on BM MSCs. Finally, using principal components analysis of the microarray data we have shown that BM MSCs constitute by membrane phenotype a specific cell population clearly distinct from that of hematopoietic cells or from other skeletal mesenchymal population (periosteal cells or synovial fibroblasts).

These data indicate that BM MSCs can be discriminated from other hematopoietic or mesenchymal populations, provided that a large number of transcripts and proteins are studied. Conversely, the use of few antigens, as preconized for example by the ISCT [11], can only indicate what these cells are not, without positively specifying them.

3.3.5 Nature and Isolation of the Native Cells

Many of the membrane proteins expressed by culture-amplified cells are modulated according to the culture conditions. Some antigens appear or increase with time in culture while others decline. In addition, some are easily inducible, being up-regulated by trace amounts of inflammatory cytokines. However, a few antigens are present at culture inception and maintained throughout the culture. We have used these antigens to isolate native MSCs from BM mono-nucleated cells [21]. Antibodies against CD73, CD130, CD146 and CD200 make it possible to isolate a very small population of cells (less than 0.5% of the mono-nucleated cells) highly enriched in CFU-Fs (enrichment factor $\geq$ 100). Presently about a dozen of antibodies, including some directed against carbohydrate moieties, are available and may be used individually or, at best, in combination to isolate BM MSCs (for a review, see [22]). Remarkably, most of the antibodies do not allow the isolation of precursors with more restricted potential than MSCs, which questions the existence of a stem cell hierarchy in the mesenchymal system.

Some of the transcripts characterizing culture-amplified BM MSCs are coding for proteins expressed by mural cells of the vasculature (*ENG*, *TEK*, *MCAM*, *VCAM1*, *DCBLD2*, *EDNRA*, *ETLD2*). Moreover, CD146$^+$ cells implanted into the immune-deficient mouse, as described above, are able to transfer the hematopoietic microenvironment [14]. After implantation these cells develop on the abluminal side of endothelial cells of incipient marrow sinuses and extend cytoplasmic processes within the marrow logette where they are in intimate contact with hematopoietic cells. Contrary to CD146+ cells evidently of donor origin, the endothelial and hematopoietic cells were provided by the recipient, which clearly establishes the CD146+ cell component as the organizing center of the logette formation. These data not only confirm the hypothesis of a HSC vascular niche but also indicate that the critical stromal component of this niche is not the endothelial cell, but an abluminal pericyte-like cells. Pericytes belong to the family of vascular smooth muscle cells and possess a contractile apparatus, which suggests that they may play a role not only in HSC maintenance but also in HSC egress from the marrow logette to the circulating blood.

3.4 Conclusions and Perspectives

Taken together, the present data allow to define BM MSCs as *bona fide* stem cells constituting a specific bone marrow population with dual properties: (i) serving as precursor for the skeletal connective tissue lineages and (ii) mandatory for the maintenance of stemness in HSCs. The two types of stem cells existing in the bone marrow of mammals, the MSCs and the HSCs, are therefore two interacting populations. A subpopulation of mural pericyte-like cells constitutes the *in vivo* reservoir of MSCs and provides the stromal component of the vascular niche. MSCs differentiating into a specific subpopulation of osteoblasts (spindle N-cadherin$^+$ cells) form the osteoblastic niche.

Although much is known on the critical regulators of HSCs by stromal cells, much less is understood on the molecular mechanisms of MSC self-renewal and commitment. Extensive "omics" studies on culture-amplified and native cells should help discriminate the essential gene networks that account for the MSC attributes.

From a clinical viewpoint, it appears essential to devise, for amplification in culture, defined synthetic media and GMP procedures that could be completely reproduced from one team to the other. This would allow strict comparison of outcomes of clinical trials using MSC-based cell therapies. The alternative of transplanting native cells is likely to be extremely difficult to implement, due to the very low frequency of these cells.

References

[1] P. Charbord. Bone marrow mesenchymal stem cells: Historical overview and concepts. *Human Gene Therapy*, 21:1045–1056, September 2010.

[2] K.A. Moore and I.R. Lemischka. Stem cells and their niches. *Science*, 311:1880–1885, March 2006.

[3] G.B. Adams and D.T. Scadden. The hematopoietic stem cell in its place. *Nat. Immunol.*, 7:333–337, April 2006.

[4] H.G. Kopp, S.T. Avecilla, A.T. Hooper, and S. Rafii. The bone marrow vascular niche: home of HSC differentiation and mobilization. *Physiology (Bethesda)*, 20:349–356, October 2005.

[5] P. Au, J. Tam, D. Fukumura, and R.K. Jain, Bone marrow-derived mesenchymal stem cells facilitate engineering of long-lasting functional vasculature. *Blood*, 111:4551–4558, May 2008.

[6] Y. Kashiwakura, Y. Katoh, K. Tamayose, H. Konishi, N. Takaya, S. Yuhara, M. Yamada, K. Sugimoto, and H. Daida. Isolation of bone marrow stromal cell-derived smooth muscle cells by a human SM22alpha promoter: in vitro differentiation of putative smooth muscle progenitor cells of bone marrow. *Circulation*, 107:2078–2081, April 2003.

[7] B. Hegner, M. Lange, A. Kusch, K. Essin, O. Sezer, E. Schulze-Lohoff, F.C. Luft, M. Gollasch, and D. Dragun. mTOR regulates vascular smooth muscle cell differentiation from human bone marrow-derived mesenchymal progenitors. *Arteriosclerosis, Thrombosis, and Vascular Biology*, 29, 232–238, February 2009.

[8] K. Kurpinski, H. Lam, J. Chu, A. Wang, A. Kim, E. Tsay, S. Agrawal, D.V. Schaffer, and S. Li. Transforming growth factor-beta and notch signaling mediate stem cell differentiation into smooth muscle cells. *Stem Cells*, 28:734–742, April 2010.

[9] B. Delorme, J. Ringe, C. Pontikoglou, J. Gaillard, A. Langonne, L. Sensebe, D. Noel, C. Jorgensen, T. Haupl, and P. Charbord. Specific lineage-priming of bone marrow mesenchymal stem cells provides the molecular framework for their plasticity. *Stem Cells*, 27:1142–1151, February 2009.

[10] M.R. Kim, E.S. Jeon, Y.M. Kim, J.S. Lee, and J.H. Kim. Thromboxane A2 induces differentiation of human mesenchymal stem cells to smooth muscle-like cells. *Stem Cells*, October 2008.

[11] E.M. Horwitz, K. Le Blanc, M. Dominici, I. Mueller, I. Slaper-Cortenbach, F.C. Marini, R.J. Deans, D.S. Krause, and A. Keating. Clarification of the nomenclature for MSC: The International Society for Cellular Therapy position statement. *Cytotherapy*, 7:393–395, 2005.

[12] P. Charbord, E. Livne, G. Gross, T. Haupl, N.M. Neves, P. Marie, P. Bianco, and C. Jorgensen. Human bone marrow mesenchymal stem cells: A systematic reappraisal via the genostem experience. *Stem Cell Reviews*, 7:32–42, March 2011.

[13] B. Delorme and P. Charbord. Culture and characterization of human bone marrow mesenchymal stem cells. *Methods Mol. Med.*, 140:67–81, 2007.

[14] B. Sacchetti, A. Funari, S. Michienzi, S. Di Cesare, S. Piersanti, I. Saggio, E. Tagliafico, S. Ferrari, P.G. Robey, M. Riminucci, and P. Bianco, Self-renewing osteoprogenitors in bone marrow sinusoids can organize a hematopoietic microenvironment. *Cell*, 131:324–336, October 2007.

[15] S. Mendez-Ferrer, T.V. Michurina, F. Ferraro, A.R. Mazloom, B.D. Macarthur, S.A. Lira, D.T. Scadden, A. Ma'ayan, G.N. Enikolopov, and P.S. Frenette. Mesenchymal and haematopoietic stem cells form a unique bone marrow niche. *Nature*, 466:829–834, August 2010.

[16] K. Takahashi, K. Tanabe, M. Ohnuki, M. Narita, T. Ichisaka, K. Tomoda, and S. Yamanaka. Induction of pluripotent stem cells from adult human fibroblasts by defined factors. *Cell*, 131:861–872, November 2007.

[17] T. Enver, M. Pera, C. Peterson, and P.W. Andrews. Stem cell states, fates, and the rules of attraction. *Cell Stem Cell*, 4:387–397, May 2009.

[18] Y. Takashima, T. Era, K. Nakao, S. Kondo, M. Kasuga, A. G. Smith, and S. Nishikawa. Neuroepithelial cells supply an initial transient wave of MSC differentiation. *Cell*, 129:1377–1388, June 2007.

[19] S. Morikawa, Y. Mabuchi, K. Niibe, S. Suzuki, N. Nagoshi, T. Sunabori, S. Shimmura, Y. Nagai, T. Nakagawa, H. Okano, and Y. Matsuzaki. Development of mesenchymal stem cells partially originate from the neural crest. *Biochem. Biophys. Res. Commun.*, 379:1114–1119, February 2009.

[20] H.Y. Chang, J.T. Chi, S. Dudoit, C. Bondre, M. van de Rijn, D. Botstein, and P.O. Brown. Diversity, topographic differentiation, and positional memory in human fibroblasts. *Proc. Natl. Acad. Sci. USA*, 99:12877–1282, October 2002.

[21] B. Delorme, J. Ringe, N. Gallay, Y. Le Vern, D. Kerboeuf, C. Jorgensen, P. Rosset, L. Sensebe, P. Layrolle, T. Haupl, and P. Charbord. Specific plasma membrane protein phenotype of culture-amplified and native human bone marrow mesenchymal stem cells. *Blood*, 111:2631–2635, March 2008.

[22] C. Pontikoglou, B. Delorme, and P. Charbord. Human bone marrow native mesenchymal stem cells. *Regen. Med.*, 3:731–741, September 2008.

4

Molecular Characterization of Stem Cells in Breast Cancer

Gabriella Di Cola[1,2], Luigi Roncoroni[3] and Leopoldo Sarli[3]

[1]Department of Human Genetics Engineering, University of Parma, Parma, Italy; e-mail: g.dicola@mail.com
[2]EMI Group and European Clinic & Research ECR, Parma, Italy;
[3]Surgical Clinic and Surgical Therapy, University of Parma, Parma, Italy

Abstract

There is experimental evidence of the presence of hallmark stem cell characteristics, such as self-renewal, undifferentiated state, and multi-potentiality in a small number of cells in cancer mass. It has earlier been thought that the cancer develops due to the progressive accumulation of random genetic mutations. But these new observations lead to another theory of origin of cancer: that cancer starts from altered adult stem cells. The tumor is considered as an aberrant organogenesis, originated and sustained by these altered stem cells called cancer stem cells or tumor initiating cells. This model had profound theoretical and therapeutic implications. The rogue stem cells can easily evade the action of anti-cancer treatments due to the fact that they can self-renew and are protected from programmed cell death. Their presence in the body can lead to the generation of tumor locally and elsewhere (metastasis). The presence of cancer stem cells and the degree of malignancy in the breast tumor can be determined by characterizing these cells with molecular markers.

In the present study, a panel of molecular markers has been investigated to develop a molecular platform based on a set of validated markers for the detection of stem cells in the breast tumors. A set of known genes, representative of cells that have the ability to self-renew, has been used in the panel.

P. Di Nardo (Ed.), Adult Stem Cell Standardization, 41–53.

Characteristic adult stem cell gene expression, with particular attention to the determinants of asymmetric division, gene markers that are indicative of embryonic origin has been studied. Cells collected from breast cancer biopsies were used in the gene expression study with reverse transcription PCR, and quantitative Real Time PCR, to highlight gene expression of one or more of the following transcription factors: Nanog, Nucleostamin, Oct4, BMI-1, DPPA4, Stellar, MSI-1 NUMB, PARD6A, PARD6B, DKK1.

Keywords: stem cells, cancer stem cells, adult stem cells, anti-cancer, breast tumor, genes.

4.1 Introduction

Experimental evidence shows the presence, within the cancer mass, of a small number of cells, with a highlighted stem characteristic, such as self-renewing, undifferentiated state and multipotentiality.

This gives rise to a new vision about the origin of cancer: instead of considering that tumors originate from the progressive accumulation of random genetic mutations, it is believed that cancer starts from a changed adult stem cell.

The tumor is considered as an aberrant organogenesis, originated and sustained by these changed stem cells, called cancer stem cells or tumor initiating cells. This model has profound theoretical and therapeutic implications. In fact, considering the ability of normal stem cells to self-renewing and to be protected by the process of programmed death, it is clear that cancer stem cells have all the characteristics to resist the working of anti-cancer treatment and it is also clear that their persistance within the body make them able to regenerate tumors locally or elsewhere (metastasis). The characterization of these cells at a molecular level is due to studies with molecular markers, and it is important to determine the actual presence of cancer stem cells in breast tumors and to see their degrees of malignancy. It has been demonstrated that the characterization with molecular markers as a prognostic indicator of cancer is more reliable than the more traditional methods.

In the specific case of breast cancer a malignant tumor originates from epithelial cells from each of the three germinal leaflets. There are different types of breast cancer even if the most frequent originate from glandular cells of the lobules or from the ducts of walls.

In human breast cancers, tumorigenic breast cancer stem cells have a phenotype: CD44+, CD24-, ESA+. These tumorigenic cells show a common phenotype and metabolic pathway as that of normal stem cells. This suggest that those cells have a common progenitor that is the target for malignant transformation and that transformed cells can function as cancer stem cells which drive tumor growth.

4.2 Aim of the Work and Genes

Despite the increasing knowledge about research and treatment of metastatic breast cancer, mortality associated with this disease remains high because of the phenomena of resistance to treatment.

The aim of this work is the creation of a molecular panel to begin the characterization of gene expression of cancer stem cells that show molecular markers ESA and CD44. Hence, this study does not discuss the detection of cancer stem cells in breast tumors.

To create this panel a set of known genes has been used, representative of cells with the ability to self regenerate in healthy tissue and stem cells: Nanog, Nucleostemin, Oct4, BMI-1, DPPA4, Stellar, MSI-1 NUMB, PARD6A, PARD6B, DKK1. Those genes are characteristic of adult stem cells, with a particular relevance to determinants of asymmetric division and markers which are indicative of embryonic origin.

Nanog is localized on the short arm of chromosome 12. It codifies for a transcription factor expressed in embryonic stem cells and is involved in the proliferation and self-renewing of the inner cell mass. If over-expressed it promotes proliferation.

Nucleostemin or GNL3 is localized on the short arm of chromosome 3. It codifies for a nucleolar protein which is a regulator of cell stem and cancer cells' growth and proliferation. It may be involved in tumerogenesis. OCT4 is localized on the short arm of chromosome 6 and codifies for an octamer-binding transcription factor. It forms a trimeric complex with SOX2 on DNA and regulates the expression of genes involved in the embryonic development. BMI-1 is localized on the short arm of chromosome 10 and encodes for polycomb ring-finger protein involved in self-renewal of cells.

DPPA4 is localized on the long arm of chromosome 3 and encodes for a development plurypotency-associated protein. Stellar is localized on the long arm of chromosome 14 and encodes for a protein expressed in embryonic stem cells. It could have a role in the maintenance of cellular plurypotency. MSI-1 is localized on the long arm of chromosome 12 and encodes for a RNA-binding protein with a central role in post-transcriptional gene regulation. NUMB is localized on the long arm of chromosome 14 and encodes for a protein involved in asymmetrical cell division and in cells' fate specification. PARD6A is localized on the long arm of chromosome 16 and encodes for a cell membrane protein involved in asymmetrical cell division and cell polarization processes. PARD6B is localized on the long arm of chromosome 20 and encodes for a cytoplasmatic protein involved in asymmetrical cell division and cell polarization processes. DKKL-1 is localized on the long arm of chromosome 10 and encodes for a secreted protein involved in embryonic development.

The expression of those genes has been analyzed in tumor cells of breast cancer and haematopoietic stem cells are used as control.

4.3 Materials and Methods

The selected cells have been collected in breast cancer biopsies of 25 patients undergoing mastectomy. Cells were selected that express the marker of staminality for the RNA extraction and the expression analysis of considered genes. At the same time, hematopoietic stem cells from peripheral blood from the same patients were selected as a means of control.

After crumbling the byopsy, they were treated with collagenase III which contains different proteolitic enzymes. The cells were prepared for the culture to obtain a significative cell growth using five different mediums: Medium 1 (RPMI 1640, FBS 10%, Streptomicin/Penicillin 1%, Glucose 2 mM 0.3 g/l, HEPES 25 mM 5.95 g/l), Medium 2 (RPMI 1640, FBS 5%, Insulin 5μg/ml, EGF 10 ng/ml, Streptomicin/Penicillin 1%), Medium 3 (MI 1640, EGF 20 ng/ml, bFGF 20 ng/ml, L-glucosio 0.5 mM, FBS 2%), Medium 4 (Chang C), Medium 5 (Amniochrome plus).

After one week of culture the second medium was chosen (RPMI, FBS 5%, insulin, EGF, streptomycin/penicillin), hence we continued the analysis only with these cells.

From this cell culture, by MACS, the class that expresses the marker of staminality CD44+, CD24- and ESA+ was selected (Figure 4.1).

4.3.1 Selection

For the selection of cells we have used the MACS system (Figure 4.2) where the sample is incubated with magnetic beads covered by antibodies against antigen. The solution is transferred in a column subjected to a strong magnetic field. For an antigen recognized by an antibody it is possible to do two different types of separation, positive and negative.

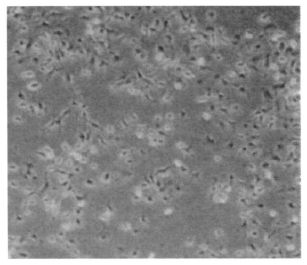

Figure 4.1 Culture of cells from breast cancer mass.

In the negative selection by MACS, an antibody is directed against an antigen which is not present in the cells concerned, so our cells are in the eluted fraction. In this way we obtain a depletion of cells that express CD protein where antibodies are directed.

In the positive selection an antibody is directed against an antigen expressed by the cells' population. The selection by MACS is divided in 3 steps: after the depletion of CD24 cells, there is a positive selection of ESA cells and CD44 cells.

Selected cells are used for RNA extraction and analysis in Real Time PCR. To compare the expression of selected genes as markers a population of hematopoietic stem cells has been used, separated by the Ficoll method, from periferic blood. Those cells express the antigen CD34 that is used for identification and enrichment.

4.3.2 RNA Extraction

RNA extraction is obtained with TRIzol reagent and it was used for selected cells through immunomagnetic separation. To the pellet, obtained after centrifugation, is added the solution of the extraction according to Chomczynsky and Sacchi's method. The sample is incubated with TRIzol reagent for 5 to 15 minutes at room temperature, to provide complete homogenization. During this step TRIzol maintains RNA molecule integrity. The addition of chloro-

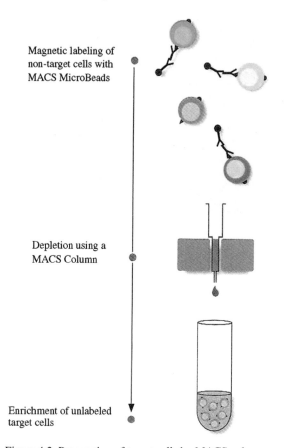

Magnetic labeling of
non-target cells with
MACS MicroBeads

Depletion using a
MACS Column

Enrichment of unlabeled
target cells

Figure 4.2 Preparation of target cells by MACS column.

form followed by centrifugation, divides the solution into three phases: one which contains phenol-chloroform and DNA, an interphase with proteins and an aqueous phase which contains RNA. RNA is precipitate through isopropyl alcohol and then centrifuged. After the RNA extraction it is essential to effectuate a treatment with DNAsi of RNA extracted. This provides an adjunct of the enzyme to degrade DNA. Then the RNA is also purified with columns of Kit RNeasy Mini of QIAGEN to prevent RNA degradation.

4.3.2.1 CD34+ Ccells Expression
To compare the expression of genes selected as a marker of staminality we used haematopoietic stem cells. These cells have a self-renewal and differen-

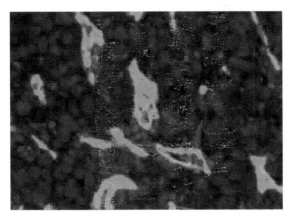

Figure 4.3 CD34+.

tiation property and give rise to all blood cell lines. Blood stem cells express the CD34 antigenic molecule, used to identify and to enrich the collection of stem cells from bone marrow or peripheral blood suspensions. From Southern blot analysis of hybrid panel human/mice it was possible to localize CD34+ gene on chromosome 1q32.

4.3.2.2 CD34+ Cells' Separation with the Ficoll Method
CC34+ cells, used to obtain control RNA for this study, have been collected from samples of peripheral blood through the Ficoll method, followed by immunomagnetic separation. This method is based on isopicnica separation and is used to separate components of whole blood, from both large and small volumes. The protocol prescribes the use of peripheral blood defibrinated and collected in test tubes with anticoagulant. Blood is mixed in the ratio 1:2 with a PBS 1 X-EDTA 2 mM solution. The mix obtained is stratified on half volume of Ficoll-Paque Plus. We have used peripheral blood of three patients, with their informed consents and for each sample we have set up two test tubes of 15 ml. After centrifugation and wash steps our cells can be conserved at 4°C for one day or submitted to imunomagnetic marking after centrifugation.

4.3.2.3 Immunomagnetic Marking of CD34+ Cells
A positive selection was effectuated with the Milteny Biotec method. Cells obtained with the Ficoll method were centrifuged and the pellet was resuspended in 300 μl of FcR Blocking Reagent and 100 μl of anti CD34+ antibodies.

The mix was incubated for 30 minutes at 4°C. After this we carried out a MACS Separation and then a RNA extraction with TRIZOL method as described below.

To verify the integrity of the RNA molecule, so for a qualitative analysis, we performed a denaturant gel electrophoresis and analysis of RNA extracted using Agilent Bioanalyzer 2100 instrument.

To obtain a quantitative analysis we performed a determination of the absorption of ultraviolet light using a spectrophotometer. For each sample, Agilent Bioanalyzer 2100 provided an electropherogram as a result and a virtual image of a typical agars gel as follows.

4.3.3 Real Time PCR

The cDNA was obtained with RevertAid First Strand cDNA Syntesis Kit and used for amplification in Real Time PCR. This kit use a genetically modified reverse transcriptase, RevertAid M-MuLV RT, with RNasi H's low activity. We used Random Hexamer primers which have the advantage of amplifying the total RNA but the disadvantage of amplifying mainly ribosomal RNA. The reaction occurs through a series of steps and the sample is incubated at various temperatures. The enzyme works in a range of temperatures between 42 and 50°C and the integrity of nucleic acid molecule is preserved by a inhibitor of recombinant RNasi, RiboLock RNasi Inhibitor, that prevent also degradation of RNA at high temperatures. RNA is placed in a test tube together with primers and water and incubated at 70°C for 5 minutes to provide the annealing of primers. The test tube is placed in ice and then the reaction buffer, the RiboLock Ribonuclease Inhibitor (20 U/μl) and a mix of single nucleotides (dNTP 10 mM) should be added.

For each PCR we used 1.5 μl of cDNA and added the Platinum SYBR Green qPCR Super mix-UDG reagent with ROX. The mix has a concentration of 2X and contains Platinum Taq DNA polymerase, SYBR Green I Dye, Tris-HCl, KCl, 6 mM MgCl2, 400 μM dGTP, 400 μM dATP, 400 μM dCTP, 400 μM dUTP, uracil DNA glycosilase (UDG) and 1 μM ROX Reference Dye. The mix automatically combines the technology of "hot-strat" Platinum Taq DNA polymerase with the technology that prevents carryover with UDG and the fluorescent dye SYBR Green to obtain an excellent sensibility to quantify target sequences.

The genes examined are, as mentioned before, characteristic of adult stem cells, and particularly those which determine asymmetrical division and marker gene that denote embryonic-histological origin of presumptive Cancer Stem Cells: Nanog, Nucleostemin, Oct4, BMI-1, DPPA4, Stellar, MSI-1 NUMB, PARD6A, PARD6B, DKK1.

4.4 Results

4.4.1 Expression of Genes

According to the results obtained, Nucleostemin is lightly under-expressed in cancer cells compared to stem cells control CD34+. Comparing Ct values (fluorescence's cycle threshold) of his expression with those of housekeeping gene 18 S in tumor tissues and CD34+ cells in the same sample, the expression of this gene is similar to the expression of genes constitutively expressed. Hence, this gene has a high transcription.

Nanog is more under-expressed in tumor cells than CD34+ stem cells. Comparing Ct values of his expression with those of 18 S we observed that in tumor tissues the gene's signal is expressed six cycles after the constitutive gene, probably because of a limited transcription. At the same time comparison in CD34+ cells shows a parallelism in fluorescence signal of Nanog gene with that of 18 S gene showing a similar transcription of the housekeeping gene.

OCT4 is over-expressed in cancer cells compared to CD34+ control stem cells. In cancer tissues the OCT4 signal is expressed seven cycles after the constitutive gene's signal, so there is a limited transcription. In CD34+ cells fluorescence shows that amplification of these genes is expressed at cycle 39 to underline no expression in these cells.

BMI-1 is lightly under-expressed in cancer cells compared to Cd34+ control stem cells. In cancer tissues the fluorescence signal is expressed 11 cycles after the constitutive gene's signal, so we can assume that there is a more limited transcription than with the OCT4 gene. Also in CD34+ cells fluorescence denotes a limited expression.

The NUMB gene is hardly under-expressed in cancer cells compared to Cd34+ control stem cells. In cancer tissues there is a fluorescence signal 23 cycles after the constitutive gene's signal. The expression level of this gene is low also in CD34+ cells. Probably it varies with different cancer stages but in this study we have not taken this into account.

The PARD6A gene's expression was not detected by instruments so it was impossible to compare ΔCt values and to obtain quantitative information on cancer tissues' expression compared to control cells. We have detected Ct

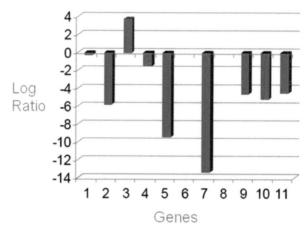

Figure 4.4 Genes' expression: 1 – Nucleostemin, 2 – Nanog, 3 – OCT4, 4 – BMI-1, 5 – NUMB, 6 – PARD6A, 7– MSI-1, 8 – Stellar, 9 – DPPA4, 10 – DKK-1, 11 – PARD6B.

values of cancer tissues and their expression 11 cycles thereafter to denote a limited expression, similar to the BMI-1 gene.

The MSI-1 gene is hardly under-expressed in cancer cells compared to CD34+ control stem cells. Ct values of his expression in cancer tissues are expressed 15 cycles after the constitutive gene's signal. Also for this gene we can think of a different expression depending on different cancer stages of the patients. We should take into consideration that the MSI-1 expression in CD34+ cells is the same as the constitutive gene's expression.

Also for the Stellar gene we had no results of CD34+ cells' expression and Ct values in cancer tissues changes between 3 to 20 cycles over 18 S gene denoting a strong variability of expression from patient to patient.

The DPPA4 gene is under-expressed in tumor cells compared to CD34+ control stem cells. Ct values of his expression in tumor tissue are expressed 11 cycles after the constitutive gene's signal, so there is a limited transcription.

PARD6A is under-expressed compared to CD34+ cells.

The DKK-1 gene is under-expressed compared to CD34+ cells. Ct values of its expression in tumor tissues are expressed 15 cycles after the constitutive gene's signal. So, there is a limit transcription, but in three patients, including patient 4, there is a similar expression with CD34+ control cells.

4.5 Conclusions

In this study we tried to create a molecular panel from known genes, to characterize gene expression of breast cancer stem cells. Results were compared with expression of the same genes in hematopoietic stem cells control CD34+.

For each of the 25 patients 12 reactions were made: 11 for marker genes and 1 for 18 S (housekeeping).

As shown in the graphics all genes are under-expressed in cancer cells compared with control cells CD34+. The only gene which seems to have an increased expression in cancer cells compared to stem cells is OCT4, so it is the main gene to create a future molecular panel that reveals genes expression in cancer stem cells.

References

[1] M. Al-Hajj, M.S. Wicha, A. Benito-Hernandez, S.J. Morrison, M.F. Clarke. Prospective identification of tumorigenic breast cancer cells. *PNAS*, 100(7):3983–3988, 2003.

[2] I.J. Fidler, M.L. Kripke. Metastasis results from pr eexisting variant cells within a malignant tumor. *Science*, 197:893–895, 1977.

[3] P.C. Nowell. Mechanisms of tumor progression. *Cancer Res.*, 46:2203–2207, 1986.

[4] C.M. Southam, A. Brunschwig. Quantitative studies of auto transplantation of human cancer. *Cancer*, 14:971–978, 1961.

[5] J.E. Dick. Breast cancer stem cells revealed. *PNAS*, 100(7):3547–3549, 2003.

[6] T. Reya, S.J. Morrison, M.F. Clarke, I.L. Weissman. Stem cells, cancer, and cancer stem cells. *Nature*, 414:105–111, 2001.

[7] D. Rubio, J. Garcia-Castro, M.C. Martin, R. de la Fuente, J.C. Cigudosa, A.C. Lloyd, A. Bernard. Spontaneus human adult stem cell trasformation. *Cancer Res.*, 65(8):3035–3039, 2005.

[8] R. Pardal, M.F. Clarke, S.J. Morrison. Applying the principles of stem-cell biology to cancer. *Nat. Rev. Cancer*, 3:895–902, 2003.

[9] U. Veronesi, P. Boyle, A. Goldrisch, R. Orecchia, G. Viale. Breast cancer. *The Lancet*, 365:1727–1741, 2005.

[10] American Joint Committee on Cancer, Breast. In *AJCC Cancer Staging Manual*, 6th ed. Springer, New York, pp. 171-180, 2002.

[11] M. Al-Hajj, M.W. Becker, M. Wicha, I. Weissman, M.F. Clarke. Therapeutic implications of cancer stem cells. *Current Opinion in Genetics & Development*, 14:43–47, 2004.

[12] K. Mitsui, Y. Tokuzawa, H. Itoh et al. The homeoprotein Nanog is required for maintenance of pluripotency in mous eepiblast and ES cells. *Cell*, 113: 631–642, 2003.

[13] S.J. Liu, Z.W. Cai, Y.J. Liu, M.Y. Dong, L.Q. Sun, G.F. Hu, Y.Y. Wei, W.D. Lao. Role of nucleostemin in growth regulation of gastric cancer, liver cancer and other malignancies. *World J. Gastroenterol.*, 10(9):1246–1249, 2004.

[14] J. Thompson, J. Itskovitz-Eldor, S. Shapiro et al. Embryonic stem cell lines derived from human blastocysts. *Science*, 282:1145–1147, 1998.

[15] A.V. Molofsky, R. Pardal, T. Iwashita, I.K. Park, M.F. Clarke, S.J. Morrison. BMI-1 dependence distinguishes neural stem cell self-renewal from progenitor proliferation. *Nature*, 425:962–967, 2003.

[16] M.E. Valk-Lingbeek, S.W.M. Bruggeman, M. van Lohuizen. Stem cells and cancer: The polycomb connection. *Cell*, 118:409–418, 2004.

[17] A.T. Clarke, R.T. Rodriguez, M.S. Bodnar, M.J. Abeyta, M.I. Cedars, P.J.Turek, M.T. Firpo, R.A. Reijo Pera. Human Stellar, Nanog, and GDF3 genes are expressed in pluripotent cells and map to chromosome 12p13, a hotspot for teratocarcinoma. *Stem Cells*, 22:169–179, 2004.

[18] S. Liu, G. Dontu, M.S. Wicha. Mammary stem cella, self-renewal pathways, and carcinogenesis. *Breast Cancer Research*, 7(3), 86–95, 2005.

[19] G. Dontu, M.S. Wicha. Survival of mammary stem cells in suspension culture: Implication for stem cell biology and neoplasia. *Journal of Mammary Gland Biology and Neoplasia*, 10(1):75–86, 2005.

[20] P. Chomczynski, N. Sacchi. Single-step method of RNA isolation by acid guanidinium thiocyanate-phenol-chloroform extraction. *Anal. Biochem.*, 162:156–159, 1987.

[21] B. Vogelstein, D. Gillespie. Preparative and analytical purification of DNA from agarose. *PNAS*, 76(2), 615–619, 1979.

[22] L. Healy, G. May, K. Gale, F. Grosveld, M. Greaves, T. Enver. The stem cell antigen CD34 functions as a regulator of hemopoietic cell adhesion. *Proc. Natl. Acad. Sci.*, 12240–12244, 1995.

[23] A. Bøyum. Isolation of mononuclear cells and granulocytes from human blood. *Scand. J. Clin. Lab. Invest.*, 21(S 97): 77–89, 1968.

[24] D.A. Skoog, J.J. Leary. *Principles of Instrumental Analysis*, 4th ed. Saunders College Publishing, Philadelphia, 1992, pp. 68.

[25] R.P. Haugland et al. *Cyclic-Substituted Unsymmetrical Cyanine Dyes*, 1995.

[26] S.T. Yue et al. *Substituted Unsymmetrical Cyanine Dyes with Selected Permeability*, 1997.

[27] H. Zipper et al. Investigations on DNA intercalation and surface binding by SYBR Green I, its structure determination and methodological implications. *Nucleic Acids Res.*, 32(12):103, 2004.

[28] Q. Chou, M. Russel, D. Birch, J. Raymond, W. Bloch. Prevention of pre-PCR mispriming and primer dimerization improves low-copy-number amplifications. *Nucl. Acids Res.*, 20:1717, 1992.

[29] D.J. Sharkey, E.R. Scalice, K.G. Christy, S.M. Atwood, J.L. Daiss. Antibodies as thermolabile switches: High temperature triggering for the polymerase chain reaction. *BioTechnology*, 12:506, 1994.

[30] M. Longo, M. Berninger, J. Hartley. Use of uracil DNA glycosylase to control carry-over contamination in polymerase chain reactions. *Gene*, 93:125, 1990.

[31] T. Lindahl, S. Ljungquist, W. Siegert, B. Nyberg, B. Sperens. DNA N-glycosidases: Properties of uracil – DNA glycosidase from Escherichia coli. *J. Biol. Chem.*, 252:3286, 1977.

[32] C.T. Wittwer, M.G. Herrmann, A.A. Moss, R.P. Rasmussen. Continuous fluorescence monitoring of rapid cycle DNA amplification. *BioTechniques*, 22:130–138, 1997.

[33] T. Ishiguro, J. Saitoh, H. Yawata, H. Yamagishi, S. Iwasaki, Y. Mitoma. Homogeneous quantitative assay of hepatitis C virus RNA by Polymerase chain reaction in the presence of a fluorescent intercalater. *Anal. Biochem.*, 229:207, 1995.

5

Foreign Body Response to Subcutaneously Implanted Scaffolds for Cardiac Tissue Engineering

Claudia Serradifalco[1], Luigi Rizzuto[1], Angela De Luca[1],
Antonella Marino Gammazza[1], Patrizia Di Marco[2],
Giovanni Cassata[2], Roberto Puleio[2], Lucia Verin[3], Antonella Motta[3],
Annalisa Guercio[2], Giovanni Zummo[1] and Valentina Di Felice[1]

[1]*Dipartimento BIONEC, Università degli Studi di Palermo, Italy;*
e-mail: valentina.difelice@unipa.it
[2]*Istituto Zooprofilattico Sperimentale della Sicilia, Palermo, Italy*
[3]*BIOtech Laboratories, Università degli Studi di Trento, Italy*

Abstract

The rapid translation of preclinical cell-based therapy to restore damaged myocardium has raised questions concerning the best cell type as well as the best delivery route, and the best time of cell injection into the myocardium. Intramyocardial injection of stem cells is by far the most-used delivery technique in preclinical studies. We have recently demonstrated that c-Kit positive cardiac progenitor cells are able to organize themselves into a tissue-like cell mass in three-dimensional cultures, and with the help of an OPLA scaffold, many cells can create an organized elementary myocardium.

We assessed the reaction of the immune system to implanted synthetic scaffolds designed to deliver cardiac progenitor cells in the infarcted region of the heart.

For the synthesis of PDLLA scaffolds, the Poly (D, L lactic acid) (RESOMER® 207, MW = 252 kDa) polymer were used (6.7%) in Dicloromethane/Dimetilformamide (DCM/DMF) 70/30 (v/v). The three-

P. Di Nardo (Ed.), Adult Stem Cell Standardization, 55–62.

dimensional structure was obtained by salt-leaching, using NaCl crystals as porosity agent (NaCl < 224 μm and < 150 μm). For the synthesis of fibrinoin scaffolds, degummed silk fibres were dried and dissolved into 9.3 m LiBr water solution (20% w/v) at 65°C for 3 h. Scaffolds with different porosities, pore size, and properties were made by freeze-drying and salt-leaching. Scaffolds embedded with collagen I and cardiac progenitor cells were implanted in the subcutaneous dorsal region of athymic Nude-*Foxn1^{nu}* mice.

PDLLA and Fibroin porous scaffolds produced by salt leaching were implanted in the subcutaneous region of nude mice, and after 45 days skin biopsies demonstrated that there was a foreign body reaction to all the implanted scaffold apart from the fibroin electrospun nets.

Keywords: tissue engineering, scaffold, poly-lactic acid, fibroin, electrospinning, foreign body reaction.

5.1 Introduction

The immune system is a major issue in any organ and tissue transplantation. Since the major histocompatibility complex (MHC) was discovered in 1967, the field of transplantation has remarkably accelerated. Antibodies, antigen-presenting cells (APC), helper and cytotoxic T-cell subsets, immune surface molecules, signalling mechanisms, and cytokines are involved in the tolerance or rejection of grafts.

When we decided to use biomaterials to deliver our stem cells to the site of implantation, we had to think about the immune system and a possible foreign body reaction (FBR). This phenomenon, also known as the host response to implanted biomaterials, involves a complex cascade of immune modulators, including various cell types, soluble mediators, and cellular interactions [1]. In some cases FBR has been considered disastrous for the patients, in other cases it has been studied in depth since after the initial inflammation it starts a period of tissue regeneration. The two major problems with FBR are a fibrotic capsule and the promotion of catabolic enzymes and reactive intermediates. In FBR we usually observe activated macrophages and giant cells, which derive from the fusion of macrophages. The presence of giant cells is driven by the release of interleukin-4 (IL-4) and IL-13 in the site of inflammation.

Many cells have been used in the past to regenerate the cardiac tissue at least *in vitro*, but the best candidate seems to be the cardiac progenitor cell present in the myocardium and easily isolated from adult tissues [2, 3]. This precursor has been injected *in vivo* both subcutaneously in SCID mice [4] and in the cardiovascular system of immunodeficient rats [5], generating partially differentiated cardiomyocytes.

One of the possible mechanisms to deliver as few stem cells as possible and in a limited area of injection is to culture these cells into biodegradable scaffolds *in vitro*, and to inject this scaffold into the desired area. The main problem with the bio-artificial materials used until now is that they induce FBR. Here we show different bio-artificial materials used and their reaction with the host immune system.

5.2 Materials and Methods

5.2.1 Scaffold Production

5.2.1.1 Becton Dickinson (BD) Three-Dimensional Open-Pore Poly-Lactic Acid (OPLA)® Scaffold

The 3D OPLA® scaffold is a synthetic polymer synthesized from D, D-L, L polylactic acid.

5.2.1.2 P(*d, l*)LA Scaffold

P(*d, l*)LA scaffolds were produced by salt leaching, using a porous agent between 150 and 224 μm. The polymer was dissolved in dichloromethane/dimethylformamide solvent (70:30 v/v) and then mixed with a sufficient amount of sodium chloride salt. The resultant mixture was dried under a chemical hood and, in order to eliminate the residual solvent and salt, subjected to continuous washes in water for three days. Finally the sponges were frozen at $-20°C$ and lyophilized for 2 days.

5.2.1.3 Silk Fibroin Scaffold

Bombyx mori polyhybrid cross silkworms' white cocoons selected by Centro Sperimentale di Gelsibachicoltura, in Como, Italy, were used. Cocoons were degummed by boiling twice in a water solution of sodium bicarbonate (1.1 and 0.4 g/L, respectively), in order to remove sericins. Degummed silk fibres were treated as previously described [6].

Three-dimensional silk fibroin scaffolds were created by freeze-drying (FD). To prepare samples, the fibroin water solution was kept at $-20°C$ for 2 hours and then dried in a freeze-dryer. Finally, samples were treated in water vapour under vacuum for 24 hours and dried in air at room temperature [6].

Silk fibroin nets were created from silk powder after freeze-drying. The fibroin powder was dissolved in 99% formic acid (Carlo Erba) at two different concentrations, 10 and 12%, by weight and the solutions were electrospun on a metal plate at 21 kV, feed rate 0.005 ml/min, target distance 7 cm. A net is made by electrospinning a solution containing 10% wt of fibroin solution in formic acid [7].

5.2.1.4 Animal Model

Scaffolds were injected in athymic Nude-*Foxn1^{nu}* mice (Harlan Laboratories). All experimental protocols used in this study were approved by the Animal Care and Use Committee of the "Ministero della Sanità".

5.2.1.5 Scaffold Microinjection

Each type of scaffold was injected in the dorsal subcutaneous region of athymic Nude-*Foxn1^{nu}* mice, one scaffold per mouse, and kept in captivity for 40 days. After that period, mice were euthanized by displacement of the first cervical vertebra, and scaffolds were excised out, fixed in a solution of Acetone-Methanol-Water (2:2:1 v/v) and embedded into paraffin. Finally, embedded scaffolds were cut into 5 μm sections and stained with haematoxylin/eosin.

5.3 Results

To investigate the possibility to use customized P(d, l)LA and fibroin scaffolds in cardiac tissue engineering we tested the immunogenic properties of porous and partially oriented scaffolds prepared with different protocols, by injecting small pieces embedded in rat collagen I in the dorsal region of athymic Nude-*Foxn1^{nu}* mice. We used collagen I as the embedding medium, because cardiac stem cells showed a higher degree of cardiac differentiation in *in vitro* three-dimensional cultures as previously described [8].

As shown in Figure 5.1, all scaffolds prepared with the method of salt leaching, both P(d, l)LA and fibroin, gave a FBR similar to the BD OPLA® scaffold previously used in our studies [8]. This FBR was characterized by the activation of macrophages with the formation of giant cells, neovascularization and encapsulation. Only fibroin partially oriented sheets did not induce a FBR, as shown in the picture.

Cells from the dermis migrated into the pores of porous scaffolds, while the same fibroblasts could not migrate between the oriented fibres of the partially oriented nets.

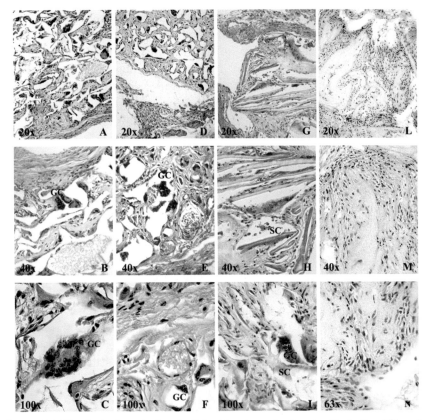

Figure 5.1 Histological analysis of skin biopsies of athymic Nude-*Foxn1^{nu}* mice. Haematoxylin/eosin stained sections: BD OPLA scaffold (A–C), porous P(*d*, *l*)LA scaffold (D–F), silk fibroin porous scaffold (G–I) and silk fibroin partial oriented nets (L–N). GC: Giant Cells; SC: Scaffold.

There are no differences among scaffolds with a different porosity.

5.4 Discussion

Cardiac stem cells are considered progenitor cells capable of differentiating into cardiomyocytes at least *in vitro* [8, 9]. To organize true sarcomeres probably they need oriented fibres as previously shown by the research group of Paolo Di Nardo [9], although we have demonstrated that leaving these cells in culture for 21 days with 20% FBS they differentiate and reach a higher degree of differentiation [10].

In the present study we wanted to investigate the immunogenic properties of P(d, l)LA and fibroin scaffolds, both porous and partially oriented, to find the best biomaterial where to let our cardiac stem cells grow before any application into a beating heart. A FBR is common for poly-lactic acid biomaterials used in orthopaedic surgery, and it is not always considered a limitation but a starting point for tissue regeneration [11]. Anyway, in our study a FBR is undesired, because it would destroy our stem cells before any differentiation may occur.

Fibroin scaffolds derive from silk and have been used to obtain structures well tolerated by the host immune system. They are mechanically robust while controlling the hydrodynamic environment, which is critical in supporting cell growth and tissue engineering [12]. The pore size may increase the gene expression of $\alpha5$, $\beta1$, and $\beta3$ integrin subunits and tissue generation, indicating that the architectural in addition to the mechanical properties of the scaffold itself are important [7].

In our study only fibroin partially oriented fibres did not induce a FBR. Both P(d, l)LA and silk fibroin porous scaffolds synthesized by salt leaching gave a FBR, characterized by giant cells, a fibrous capsule and neovascularization. In partially oriented nets, cells from the dermis colonized the scaffold without entering the structure itself, because the spaces between fibres are of nanometric size. Only fibroblasts colonized the surface of the scaffold, and there was neither fibrous capsule nor neovascularization.

The only step in the material preparation common to P(d, l)LA and three-dimensional silk fibroin scaffolds is the salt leaching method. Probably this method may create an immunogenic film over the surface of the scaffold.

A FBR for fibroin scaffolds has been reported only once by Dal Pra and colleagues [13], where they studied the immunogenicity of disorganized degummed fibroin fibres. They reported a mild FBR.

Our study is important because we found a biodegradable non-immunogenic biomaterial which can be used to deliver our cardiac stem cells to the site of injection, and because it is the first time that an FBR has been correlated to the material preparation method.

Our aim is to find a proper and standardizable protocol for scaffold production which, avoiding salt-leaching, ensures the non-immunogenicity of the implanted construct. This step is very important in the standardization of stem cell protocols for clinical application since, as it is known, the consequences of FBR to the material used can be dangerous, so, as we would like to use this scaffold to allow tissue regeneration, we have to guarantee the absence of FBR in the site of implantation. Finally, the new biomaterial produced will be able to provide the correct three-dimensional structure for stem cell differentiation that will promote myocardial regeneration.

References

[1] D.M. Higgins, R.J. Basaraba, A.C. Hohnbaum, E.J. Lee, D.W. Grainger, and M. Gonzalez-Juarrero. Localized immunosuppressive environment in the foreign body response to implanted biomaterials. *American Journal of Pathology*, 175(1):161–170, 2009.

[2] V. Di Felice, A. De Luca, M.L. Colorito, A. Montalbano, N.M. Ardizzone, F. Macaluso, A. Marino Gammazza, F. Cappello, and G. Zummo. Cardiac stem cell research: An elephant in the room?, *Anat. Rec.*, 292(3):449–454, 2009.

[3] P. Di Nardo, G. Forte, A. Ahluwalia, and M. Minieri. Cardiac progenitor cells: Potency and control. *Journal of Cellular Physiology*, 224(3):590–600, 2010.

[4] E. Messina, L. De Angelis, G. Frati, S. Morrone, S. Chimenti, F. Fiordaliso, M. Salio, M. Battaglia, M.V. Latronico, M. Coletta, E. Vivarelli, L. Frati, G. Cossu, and A. Giacomello. Isolation and expansion of adult cardiac stem cells from human and murine heart. *Circulation Research*, 95(9):911–921, 2004.

[5] A.P. Beltrami, L. Barlucchi, D. Torella, M. Baker, F. Limana, S. Chimenti, H. Kasahara, M. Rota, E. Musso, K. Urbanek, A. Leri, J. Kajstura, B. Nadal-Ginard, and P. Anversa. Adult cardiac stem cells are multipotent and support myocardial regeneration. *Cell*, 114(6):763–776, 2003.

[6] Y. Wang, E. Bella, C.S. Lee, C. Migliaresi, L. Pelcastre, Z. Schwartz, B.D. Boyan, and A. Motta. The synergistic effects of 3-D porous silk fibroin matrix scaffold properties and hydrodynamic environment in cartilage tissue regeneration. *Biomaterials*, 31(17), 4672–4681.

[7] B. Bondar, S. Fuchs, A. Motta, C. Migliaresi, and C.J. Kirkpatrick. Functionality of endothelial cells on silk fibroin nets: Comparative study of micro- and nanometric fibre size. *Biomaterials*, 29(5):561–572, 2008.

[8] V. Di Felice, N.M. Ardizzone, A. De Luca, V. Marciano, A.M. Gammazza, F. Macaluso, L. Manente, F. Cappello, A. De Luca, and G. Zummo. OPLA scaffold, collagen I, and horse serum induce an higher degree of myogenic differentiation of adult rat cardiac stem cells. *Journal of Cellular Physiology*, 221(3):729–739, 2009.

[9] G. Forte, F. Carotenuto, F. Pagliari, S. Pagliari, P. Cossa, R. Fiaccavento, A. Ahluwalia, G. Vozzi, B. Vinci, A. Serafino, A. Rinaldi, E. Traversa, L. Carosella, M. Minieri, and P. Di Nardo. Criticality of the biological and physical stimuli array inducing resident cardiac stem cell determination. *Stem Cells*, 26:2093–2103, 2008.

[10] V. Di Felice, A. De Luca, C. Serradifalco, P. Di Marco, L. Verin, A. Motta, A. Guercio, and G. Zummo. Adult stem cells, scaffolds for *in vivo* and *in vitro* myocardial tissue engineering. *Italian Journal of Anatomy and Embryology*, 115(1/2):65–69, 2010.

[11] J.M. Anderson, A. Rodriguez, and D.T. Chang. Foreign body reaction to biomaterials. *Seminars in Immunology*, 20(2):86–100, 2008.

[12] Y. Wang, U.J. Kim, D.J. Blasioli, H.J. Kim, and D.L. Kaplan. *in vitro* cartilage tissue engineering with 3D porous aqueous-derived silk scaffolds and mesenchymal stem cells. *Biomaterials*, 26(34):7082–7094, 2005.

[13] I. Dal Pra, G. Freddi, J. Minic, A. Chiarini, and U. Armato. De novo engineering of reticular connective tissue *in vivo* by silk fibroin nonwoven materials. *Biomaterials*, 26(14):1987–1999, 2005.

6

The Ethics of Stem Cell Research Publication

Graham Parker

The Carman and Ann Adams Department of Pediatrics, Wayne State University, School of Medicine, Children's Hospital of Michigan, Detroit, USA; e-mail: gparker@med.wayne.edu

Abstract

The scientific peer review process is coming under increasing scrutiny. Stem cell research in particular and its communication have become a matter of great public interest. This chapter will touch on issues relevant to the process. These issues are neither exclusive to stem cell research, nor exhaustive of the issues that are of concern to the field. Stem cell researchers are a truly an international community, although communication is primarily in English. However, in spite of, or because of, international recognition of the importance of ethics in the research and publication process, these issues warrant further consideration.

Keywords: ethics, stem cells, misconduct, clinical translation.

6.1 Introduction

This chapter is written from the perspective of an editor who firmly believes in the peer review process and the collegiality of researchers. Unfortunately, for one reason or another, individuals can act in ways that are inconsistent with what I consider to be internationally agreed upon and understood tenets of behavior with respect to scientific publishing. Some of these issues do

P. Di Nardo (Ed.), Adult Stem Cell Standardization, 63–72.

indeed occur at the individual level, while, perhaps more interestingly, some of them appear to be associated with whole sections of a field of research, a collective miasma. Although I would contend that the vast majority of the world's researchers would be in broad agreement as to what constitutes acts of misconduct, there are areas where the consensus is less complete, and also instances, for example, in proposing preferred reviewers, where researchers from all over the world regularly act in a manner that would appear to be in obvious conflict.

6.2 Scientific Misconduct

6.2.1 Definition

The National Institutes of Health (NIH) of the United States defines scientific misconduct as: "Fabrication, falsification, or plagiarism in proposing, performing, or reviewing research, or in reporting research results [...] Research misconduct does not include honest error or differences of opinion. (42 C.F.R. 93.103)" [1]. Briefly: Fabrication is the deliberate manufacturing of a data set for dissemination; Falsification involves the deliberate manipulation of a method or withholding of data from a final data set for dissemination; Plagiarism is the use of someone else's ideas and more specifically, words, without giving appropriate credit or attribution. Of these, the last requires further qualification being the most often debated as to what instances would qualify as misconduct across discipline and culture. The cutting and pasting of text from another author's work into a paper is the most obvious example. However, appropriating a thesis – a logical sequence of points building to convey an argument – without crediting the originator is also plagiarism.

The latter may only be detected by the original author or someone extremely familiar with the work and is far less likely to be pursued in a scientific journal as a case of plagiarism. This is an example of how applicability differs from a scientific discipline to, e.g., the humanities. The most extreme action a science editor might take is to request a citation be included to the original work. The former example is the more familiar and much more likely to require action. The prevalence and ease with which text can now be appropriated into a manuscript is, however, matched only by the ease with which it can now be detected. A few years ago, such an offence might be detected by an editor or reviewer recognizing a section of text where the register or tone had changed, or, even more likely, encountering a section of fluent prose where the preceding and following paragraph had been telegraphic and incoherent. More recently, pasting of the text into an internet search engine would reveal whether the text was unique or previously published elsewhere. Now, it is common for journals to use software that compares entire manuscripts to that of previously published work.

Far from making the job of the editor easier by relieving them of the burden of detecting such infringements, this software actually creates a difficult situation. For surely, once an editor is aware of a problem with a text it is then their responsibility to discover whether what has occurred is in fact plagiarism, and if so, act on it. The detection software, legitimizing itself with its pronouncements of percentages of similarity, must be interpreted with discernment as to how the software has determined such a figure. For example, I have rejected a paper with a percentage similarity of 25%, and accepted a paper that has generated a similarity index of 37%. The software can take, e.g., eight words out of the eleven in a given sentence and determine that that makes it similar to another sentence published previously. It is. However, there are only so many ways to communicate an idea. It is the responsibility of the editor to give the author the benefit of the doubt and reasonably judge their intent. If in a paragraph there are three contiguous sentences that each use the vast majority of the same words that all come from the same previously published paper, action may be required. Far easier to discern are instances where a whole paragraph or more is taken.

Even then, the editor must contact the corresponding author – the guarantor of the integrity of the manuscript – and invite them to account for the apparent discrepancy. The guarantor of the work must be given every possible opportunity to be as frank and communicative as possible during the process. Once an editor can demonstrate that not only has there been an act that is inconsistent with normal publishing practice but that they have also been misled about the responsibility for the act, the repercussions are far more likely to become long term. If the explanation is unreasonable, or, as often happens no explanation is forthcoming, then the head of the department must be contacted and where necessary the appropriate funding agency. It is then the responsibility of the author's institution to determine how to proceed and this differs widely as a formalized process between and within different countries. The editor of the journal decides the extent of sanctions with regard to their journal. Typically, an author might be banned from submitting papers to the journal for a period of time, e.g., three years. This also varies widely across journals and disciplines.

6.2.2 How Prevalent Is It?

These are issues that are the subject of water cooler discussions rather than formal analysis and publication. Even the most panglossian of readers might be surprised to learn that between 1992 and 2005 the office of research integrity for the NIH of the United States only determined fault in 162 cases of scientific misconduct [2]. Of these, only eight were solely cases of plagiarism. Eight cases in thirteen years. The journal for which I serve as editor-in-chief, *Stem Cells and Development*, had just under 600 papers submitted for review last year. There was a case of plagiarism at least once every two months. Thus far, incidentally, none has involved a paper submitted from the United States, so perhaps the figures from the NIH are less surprising. The low number of cases reported by NIH likely reflects a number of issues. Firstly, most editors and the institutions employing the authors lack the appetite to pursue such an issue. Secondly, plagiarism is often seen as a 'lesser' crime compared to falsification or fabrication. More complicated are the cultural differences. I have had the opinion expressed to me that the authors from certain cultures consider themselves to be 'quoting the poets' by using words and phrases they admire in others work. Let me blunt. If that is the case, use inverted commas and remove the ambiguity.

6.2.3 Manuscript Submission

Authors submitting a manuscript for publication in a peer-reviewed journal are typically invited to recommend reviewers they feel are appropriately experienced to be able to evaluate the submitted work, and similarly indicate the names of reviewers they would rather not be invited to review their manuscript. The software *Stem Cells and Development* uses the labels 'preferred' and 'non-preferred', which is a slightly unfortunate terminology. It encourages the author to select someone they know who would favorably review their paper, rather than the reviewer that is the best qualified. Generally, authors recommend reviewers who are indeed suitably qualified to review their paper. However, there is a growing problem of authors listing former collaborators. Does it really need to be stated that the people one has published with before are not suitable candidates? Apparently so, as many papers submitted, and this applies just as much to papers submitted from Europe and the United States as elsewhere, list former collaborators as recommended reviewers. An author listing a former co-author as a reviewer should explain in the covering letter why the subject matter is so esoteric as to make that person one of the very few qualified to review that work. Failure to do so presents a clear conflict of interest. Of a far more insidious nature, reviewer recommendations have been received that list fellow members of advisory boards to private companies. Are these people qualified to review the paper? Yes. Does their shared involvement in a private company represent a conflict of interest? Absolutely, and the author should be divulging that relationship in the covering letter.

The rationale for listing non-preferred reviewers has similar concerns. However, the most common misuse of this list is in naming an individual simply because they are a competitor to you in the field. Authors can expect to be invited to justify the exclusion of a particular reviewer if there is no obvious rationale. An editor much appreciates a carefully worded explanation for their listing. On the other side, editors appreciate invited reviewers explaining a possible conflict, such as their current undertaking of a project that is so similar to the work in question that they feel unable to offer a fair evaluation or that access to the manuscript might unfairly influence their own work. What is rather bizarre is when reviewers request a judgment as to whether they are in conflict. Do we really need someone to tell us these things? There is no good rationale for offering an impartial review. We surely know when we are in conflict or not and know our collegiate duty to act accordingly.

6.3 Representation of Stem Cell Research

One of the advantages of working in academia is the perceived freedom to choose the topic of investigation. Our focus and the passion we generate for our work can on occasions result in our overstating the significance of achievements. When writing the lay descriptions for grant applications we make our work sound as convincingly important as possible. This can lead to unrealistically high expectations of the immediacy of effective cures and scientific breakthroughs. A related issue is the language we use to describe the cells we study.

6.3.1 How We Represent Our Work

An example of the gap between the representation of the work and the work itself can be found in the culture and expansion of hematopoietic stem cells (HSC). Journals covering HSC will regularly receive papers that indicate a refined understanding that allows the authors to improve their expansion of an original blood cell population while retaining their primitive marker expressions. So why is it that the blood banks, the reserves of donated blood used for later transfusion, are not full? The fact is we are still relatively ignorant concerning the one stem cell we know more about than any other, which continues to be the stem cell against whose properties all other stem cells are compared to and found wanting [3]. The truth is we are still dependent on those original donated blood samples and are not yet capable of expanding HSC *in vitro* for general clinical use. This is not to say advances are not being made, only that the limitations of the clinical implications of that work are not properly acknowledged.

Researchers are generally naive about how quickly and widely news of their research claims is now disseminated and discussed in websites and blogs devoted to regenerative medicine and stem cell research in general. Previously, if a journal thought a piece of work was important enough they might choose to write up a press release that would typically contain a quote from the editor hopefully responsibly indicating the importance of the work. Principal investigators have to fulfill many roles: Scientist, manager, book keeper, accountant, statistician, author, etc. But the one for which they are probably the least well prepared for is that of being an interviewee dealing with reporters. A relatively recent development is that of host academic in-stitutions releasing press releases of their own to promote their researchers' work. These releases often contain language that is far more sensationalized than a journal press release would be. In response to this, I have recently added a request to our emailed letters of acceptance to ask that authors co-ordinate their promotion efforts with our publishers. The reason for this is that although we have a responsibility to stand by the integrity of the work we publish, we also have a responsibility in our interaction with the media to be realistic in our discussion of that work within its limits. If the paper shows, for example, an increase in engraftment of a given cell into a given tissue in four out of eight impaired mice that also showed an increase in an index of recovery, that might well be of high interest to the research field. Has the condition the impaired mice are modeling been cured? No. Not yet anyway, and the public, who, in many countries are in some measure funding at least part of the research, deserve not to be misled about its significance.

6.3.2 Silent Data

This issue is by no means unique to our field. A given data set is analyzed and written up and submitted to a journal for review with a view to being published. This paper represents a good deal of hard work and substantial investment by the responsible laboratory. It also likely represents more. It represents the culmination of a series of experiments, some of which worked the first time they were performed, some of which may have had to be repeated a number of times before the data set was achieved that was submitted for review. There are numerous ways an experiment can go wrong and the reason be apparent. There are also occasions where an experiment does not achieve the expected result without an obvious reason. In both situations, the work was paid for, both in person hours and expended resources, but also in the spending of funds. The data exist, are presumably properly documented, but they are silent. They are not part of our sum knowledge of the record as pertains to that experiment.

Similarly, all researchers will recognize the experience of achieving a data set about which there is so much excitement that it is written up and submitted to a 'top tier' journal, only to have it rejected due to it being either too similar to an existing or competing paper, or being 'behind the wave' of what was the current thinking of that field. Such work would typically then be reformatted and submitted to a 'second tier' journal. However, the review process may have been so disheartening or lengthy that the work in fact never gets resubmitted for publication. Such work, too, is silent. It is not part of our sum knowledge of the record as pertains to that experiment.

Although the circumstances differ, the end result is the same. Now, if the silent data set is merely confirmatory of the results of another study, one might consider the matter moot. However, in the modern era one of the most powerful and increasingly important analytical tools we have is the meta-analysis. A meta-analysis provides an approach to combine those separate studies and, with appropriate caveats and conservative interpretation, allow inferences to be drawn that would not be possible from the individual studies [4, 5]. The most obvious use of meta-analysis is for clinical trials data, where there may be a number of sets of data from individual trials that by themselves lack sufficient statistical power to draw solid conclusions.

Modern science is too expensive for us to have data sets languishing unused. The public demand for advancements and accountability of the use of funds should be stimulus enough to make us reconsider how we treat precious data beyond our current limited view of them as localized packets of information to achieve a publication. It is our responsibility to consider novel ways of making silent data available beyond our existing publication model for wider review, analysis and subsequent re-analysis.

6.3.3 How Well Do We Represent Humanity

There is a burgeoning literature examining the heterogeneity of those cells we vaguely refer to as mesenchymal stromal or stem cells (MSC). It is interesting that papers describe differences between populations of these cells isolated from different parts of the body, given that the cells isolated from even the most restricted area of one part of the body yields a population that defies any attempt at an exact phenotypic characterization. However such work is vitally important if the much anticipated translation from laboratory phenomenon to clinical treatment is to be responsibly achieved.

There is tremendous optimism for the use of pluripotent stem cells, whether they are embryonic (ESC) or 'induced', for therapeutic use. Their immunological, proliferative and pluripotent differential fate profile excite the representation of their significance for use in medicine. Their use is anticipated to be in *in vitro* assays as well as for *in vivo* applications. However, as was recently shown in two separate publications [6, 7], the existing pluripotent cell lines that are currently being used all over the world in no way reflect the racial diversity of the human race. By genotyping single-nucleotide polymorphisms in those lines and comparing them to reference samples of known genetic background it was revealed that all but two of the most commonly used ESC lines were of European or middle eastern origin, the other two being of East Asian derivation. None of the existing ESC lines have African ancestry. The implications of these findings are clear. Any information garnered from these lines is based on lines that represent less than 12% of the world's population. The field of pharmacogenetics studies how differences in genes can influence the body's response and metabolism of different drugs. That people of different genetic background react differently to medications has huge implications for the use of stem cell assays in pharmaceutical testing. At best the data sets contain systematic bias, at worst the treatments developed may contain unanticipated adverse drug reactions and tissue rejections.

What I have yet to see considered is whether any comparable problem may exist with the somatic stem cell populations being considered for therapeutic purposes. Once thought immunologically silent, at least as immature cells, we now know that MSC have considerably complicated immunomodulatory properties [8]. As far as I know, no-one has undertaken to examine how genetic background influences the properties of these fascinating cells or subjected to them to the kind of genetic ancestry analyses discussed above for ESC.

6.4 Clinical Translation

Recently I had the privilege to be part of a panel receiving questions from the public organised as part of the 2010 World Stem Cell Summit in Detroit, USA. The inevitable question came up concerning the merits of medical tourism, travelling to foreign countries in order to receive treatments not available in the United States. It is one thing to write an editorial on the subject from one's office, quite another to be faced with a mother prepared to do anything to see her son cured of an intractable condition. I advised the mother that her specialists, regardless of whether they offered such treatments, were aware of all the options and would be able to advise her of their own opinion of whether the treatments available elsewhere were worthy of consideration. The legitimacy of the purported treatments is rightly questioned. The design of the clinical trials, the robustness and generalizability of the treatments, even the characterization and identification of the stem cells used have to be standardized globally [9]. Only with such rigor can fair assessments be made about the advisability of available procedures.

Conditions that are intractable also provide an extreme example of a current concern for which there are no easy answers. Namely, how can a researcher/clinician obtain truly impartial consent for a therapeutic trial where the patient or parent of the patient has no other options? This is not to say the efforts of those involved are not well-intentioned. However, there is a global imperative to reach a more standardized view of what constitutes a preponderance of experimental evidence that justifies the translation to the clinic and thence what is an acceptable design to perform that clinical trial. A lack of resources to perform preparatory supportive bench research but an availability of suitable patients and permissive regulatory climate cannot be condoned as justification for accelerated clinical translation.

It is probably not generally understood just how expensive it is to perform clinical trials. This has led to the performing of an increasing number of those trials in countries where the cost is considerably lower. However, the problems alluded to in Section 6.3.3 also apply to the outsourcing of clinical trials to countries whose demographic are very different to the likely target patients. This of course is a different question to that of the potential exploitation of a vulnerable subject population.

6.5 Conclusions

Just as the standardization of our methods and materials is vital to the advancement of the science in our field, so is the standardization of our publication ethics vital to the advancement of its communication and its estimation by the public. I will conclude with two questions that we must answer: Who profits most from our science? And, how can we better represent our science?

References

[1] http://www.ori.dhhs.gov/documents/42_cfr_parts_50_and_93_2005.pdf.

[2] Price, A.R. http://hdl.handle.net/2027/spo.5240451.0001.001, 2006.

[3] Parker, G.C., Anastassova-Kristeva, M., Broxmeyer, H.E., Dodge, W.H., Eisenberg, L.M., Gehling, U.M., Guenin, L.M., Huss, R., Moldovan, N.I., Rao, M.S., Srour, E.F. and Yoder, M.C.. Stem cells: Shibboleths of development. *Stem Cells and Development*, 13(6):579–584, December 2004.

[4] Thomas, R.L., Parker, G.C., Van Overmeire, B. and J.V. Aranda, J.V. A meta-analysis of ibuprofen versus indomethacin for closure of patent ductus arteriosus. *European Journal of Pediatrics*, 164(3):135–140, March 2005.

[5] Janowski, M., Walczak, P. and Date, I. Intravenous route of cell delivery for treatment of neurological disorders: A meta-analysis of preclinical results. *Stem Cells and Development*, 19(1):5–16, January 2010.

[6] Mosher, J.T., Pemberton, T.J., Harter, K., Wang, C., Buzbas, E.O., Dvorak, P., Simón, C., Morrison, S.J. and Rosenberg, N.A. Lack of population diversity in commonly used human embryonic stem-cell lines. *The New England Journal of Medicine*, 362:183–185, January 2010 (doi: 10.1056/NEJMc0910371).

[7] Laurent, L.C., Nievergelt, C.M., Lynch, C., Fakunle, E., Harness, J.V., Schmidt, U., Galat, V., Laslett, A.L., Otonkoski, T., Keirstead, H.S., Schork, A., Park, H.S. and Loring, J.F. Restricted ethnic diversity in human embryonic stem cell lines. *Nature Methods*, 7(1):6–7, January 2010.

[8] Zhao, S., Wehner, R., Bornhäuser, M., Wassmuth, R., Bachmann, M. and Schmitz, M. Immunomodulatory properties of mesenchymal stromal cells and their therapeutic consequences for immune-mediated disorders. *Stem Cells and Development*, 19(5):607–614, May 2010.

[9] Di Nardo, P. and Parker, G.C. Stem cell standardization. *Stem Cells and Development*, 20(3):375–377.

7

The Role of Purinergic Receptors in Stem Cells in Their Derived Tissues

Edda Tobiasch and Yu Zhang

University of Applied Sciences, Bonn-Rhein-Sieg, Germany;
e-mail: edda.tobiasch@fh-brs.de

Abstract

Purinergic receptors are well known for their important role in many cellular processes in a huge variety of different cell types. Several publications have demonstrated that purinoceptors are not only also expressed in all kinds of stem cells, but can regulate stem cell proliferation and differentiation, and influence their metabolism and signalling. Thus, stem cell fate might be controlled by using potent and selective purinergic receptor agonists or antagonists, which moreover could be applied to induce the desired stem cell-derived tissue cell type for future application in Regenerative Medicine. In this chapter, the current findings on purinergic receptors and their signalling in stem cells are summarized and some future aspects and key research directions are discussed.

Keywords: purinergic receptor, adenosine receptor, P1 receptor, P2X receptor, P2Y receptor, extracellular nucleotide, ATP, adenosine, embryonic stem cell, hematopoietic stem cell, mesenchymal stem cell.

7.1 Introduction

Stem cells characterized by the ability to renew themselves and differentiate into a wide range of specialized cell types, are promising tools for therapeutic

P. Di Nardo (Ed.), Adult Stem Cell Standardization, 73–98.

applications in tissue or organ reconstruction and transplantation. However, it is still unclear how to provide the correct environment for the desired differentiation into the original phenotype, because the triggering signals and the underlying signal transduction pathways have been uncovered only fragmentarily and results have been only published in scattered reports until now.

This is especially true for the family of purinergic receptors and their ligands, the extracellular nucleotides. Adenosine 5'-triphosphate (ATP) was regarded only as an energy molecule in metabolism until it was demonstrated by Burnstock in 1972 to be a signalling molecule as well, which has a large array of specific receptors on the cell membrane. Until now, nineteen extra cellular nucleotide receptors have been cloned and characterized, the so called purinoceptors [3]. Their distribution, structure and expression have been studied in a variety of cells from different tissues and organs, which revealed their important functional role in a large range of physiological processes, including immune response, angiogenesis, neural transmission, apoptosis, and embryonic development [4]. The importance of extracellular nucleotides in stem cell differentiation, proliferation, migration and apoptosis has been reported as well [5, 6]. This review will summarize the influence of extra cellular nucleotide via purinergic receptors on stem cells and their consecutive tissues.

7.2　Purinergic Receptors: What Are They?

Purinergic receptors are thought to be one of the oldest receptor families in evolutionary terms [7]. They were classified by Burnstock in 1978 [8] into two main families, the P1 receptors binding to adenosine and the P2 receptors with ATP, UTP, their breakdown products ADP and UDP being their most important agonists (Figure 7.1).

To date four adenosine or P1 receptors namely A_1, A_{2A}, A_{2B} and A_3 were cloned and characterized, all coupling to G proteins [9–11]. The seven ionotropic P2X receptors ($P2X_{1-7}$), ligand-gated ion channels, can mediate the influx and efflux of cations [12, 13] and the eight metabotropic P2Y receptors ($P2Y_1$, $P2Y_2$, $P2Y_4$, $P2Y_6$, $P2Y_{11}$, $P2Y_{12}$, $P2Y_{13}$, $P2Y_{14}$), G protein-coupled receptors [14, 15], regulate a series of intracellular signal cascades after binding their agonist [16, 17]. These receptors have been cloned and characterized as well (Table 7.1). The missing numbers in the labelling of P2 receptors are owed to non-functional or variant receptors in different species.

Table 7.1 Characteristics of purine-mediated receptors.

Receptor		Distribution in stem cells	Main distribution in tissue	Structure information n	Selective agonists	Selective antagonists
P1 (adenosine)	A_1	HSCs; BM-MSCs	heart; brain; spinal cord; testis	$G_{i/o}$ 326 aa	CCPA; CPA; CHA; CVT-510; GR79236	CPX; CPT; MRS1754; N-0840; WRC-0571
	A_{2A}	HSCs; BM-MSCs	heart; brain; lung; spleen	G_s 412 aa	ATL-146e; CVT-3146; CGS 21680; DPMA; HE-NECA	KF17837; KW 6002; SCH58261; ZM241385
	A_{2B}	Not clear	bladder; large intestine	G_s 332 aa	NECA; Bay60-6583	Alloxazine; MRE2029-F20; MRS17541; MRS 1706; PSB0788
	A_3	HSCs; BM-MSCs	heart; brain; lung; liver	$G_{i/o}$ 318 aa	IB-MECA; 2-Cl-IB-MECA; DBXRM; VT160	MRS1220; MRS1191; MRS1523; MRS1292; VUF5574
P2X (ATP)	$P2X_1$	HSCs	smooth muscle; blood vessels; platelets; cerebellum; heart	cation channel 399 aa	2-MeSATP; α, β-MeSATP; BzATP; PAPET-ATP; CTP; Ap$_5$A	TNP-ATP; MRS2159; isoPPADS; Ip$_5$I; NF023; NF279; NF449; PPNDS; Phenol red
	$P2X_2$	HSCs	smooth muscle; CNS; retina; ganglia	ion channel 471aa	ATPγS; 2-MeSATP; Ap$_4$A	PPADS; RB2; NF279; TNP-ATP
	$P2X_3$	ESCs; ATSCs; HSCs	sensory neurones; NTS; heart; cochlea	cation channel 397 aa	2-MeSATP; α, β-meATP; Ap$_5$A; PAPET-ATP; HT-AMP	A317491; TNP-ATP; isoPPADS; Ip$_5$I; NF023; RO4; RO85
	$P2X_4$	ESCs; ATSCs; HSCs	CNS; testis; colon; smooth muscle; epithelia	ion channel 388 aa	α, β-meATP; CTP; ATPγS	Phenolphthalein; BBG; TNP-ATP; Paroxetine
	$P2X_5$	ATSCs; HSCs	thymus; gut; spinal cord; heart	ion channel 444 aa	2-MeSATP; ATPγS; α, β-meATP; GTP; BzATP	PPADS; BBG
	$P2X_6$	ATSCs; HSCs	CNS; sensory neurons; epithelia	ion channel 379 aa	(does not function as homo-multimer)	–
	$P2X_7$	ATSCs; HSCs;	Pancreas; skin;	cation	2-MeSATP;	BBG;

Table 7.1 (Continued)

		BM-SMCs	bone marrow	channel 595 aa	BzATP; α, β-meATP	KN62; KN04; Oxidized ATP; RN-6189; AZ10606120; A-740003
P2Y (ATP; UTP; ADP; UDP)	P2Y$_1$	ESCs; BM-SMCs; ATSCs; HSCs	endothelia; epithelia; smooth muscle; platelets; CNS	G$_{q/11}$ 372 aa	ADPβS; 2-MeSADP; 2-MeSATP; MRS2365	A3P5PS; BzATP; MRS2179; MRS2500; MRS2279; PIT; PPADS
	P2Y$_2$	ESCs; BM-SMCs; ATSCs; HSCs	epithelia; endothelia; kidney tubules; smooth muscle	G$_{q/11}$ 376 aa	MRS2698; INS 37217; INS 365; NS365; UTP; UTPγS	ARC126313; MRS2576; PSB-716; RB2
	P2Y$_4$	ATSCs	endothelia; spleen; CNS; smooth muscle; placenta; thymus	G$_{q/11}$ and G$_i$ 365 aa	UTP; Up$_4$U; INS365; INS 37217	RB2; MRS2577; PPADS; BzATP
	P2Y$_6$	ATSCs	placenta; thymus; epithelia; heart; smooth muscle	G$_{q/11}$ 328 aa	UDPβS; UDP; α, β-methylene-UDP; INS48823; MRS2693; IDP	MRS2578; RB2; PPADS; MRS2567; MRS2575
	P2Y$_{11}$	HSCs; ATSCs	spleen; endothelia; smooth muscle; intestine	G$_{q/11}$ and G$_s$	ARC67085MX; BzATP; ADPβS; NF546	RB2; 5'- AMPS; AMPα5; NF157; NF340
	P2Y$_{12}$	ATSCs	platelets; glia; microglia	G$_{i/o}$ 342 aa	2-MeSADP; ADP; IDP	AR-C69931MX; AR-C67085; CT50547; INS50589; C1330-7; RB2; 2-MeSAMP; MRS2395
	P2Y$_{13}$	ATSCs	brain; bone marrow; spleen; lymph nodes	G$_{i/o}$ 333 aa	ADP; 2-MeSADP; ADPβS; IDP	AR-C67085; AR-C69931MX; MRS2211; 2-MeSAMP; PPADS
	P2Y$_{14}$	ATSCs; HSCs	adipose tissue; stomach; intestine; placenta	G$_{q/11}$ 338 aa	UDP-glucose; UDP-galactose; UDP-glucosamine; MRS2690	–

Source: modified after sigma-aldrich.com/ehandbook_01.2011 and Burnstock [3] and extended.

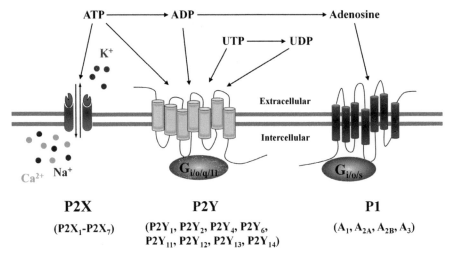

P2X

(P2X$_1$-P2X$_7$)

P2Y

(P2Y$_1$, P2Y$_2$, P2Y$_4$, P2Y$_6$,
P2Y$_{11}$, P2Y$_{12}$, P2Y$_{13}$, P2Y$_{14}$)

P1

(A$_1$, A$_{2A}$, A$_{2B}$, A$_3$)

Figure 7.1 Purinergic receptors. Purinergic receptors can be divided into P1 receptors (A$_1$, A$_{2A}$, A$_{2B}$ and A$_3$), P2X ionotropic receptors (P2X$_{1-7}$), and metabotropic P2Y receptors (P2Y$_1$, P2Y$_2$, P2Y$_4$, P2Y$_6$, P2Y$_{11}$, P2Y$_{12}$, P2Y$_{13}$, P2Y$_{14}$). P1 receptors are activated via adenosine whereas P2 receptors react to ATP and UTP and their breakdown product ADP and UDP.

7.2.1 P1 or Adenosine Receptors

P1 or adenosine receptors couple to adenylate cyclase in general. A$_{2A}$ and A$_{2B}$ belong to G$_s$ coupled proteins and are positively connected with adenylate cyclase. In contrast to this, A$_1$ and A$_3$ are negatively coupled to it via the G$_{i/o}$ protein α-subunits [18]. However, recently the human A$_{2B}$ receptor has also been found to trigger phospholipase C activity via G$_{q/11}$, and the A$_3$ receptor may oppositely interact directly with G$_s$ protein (Figure 7.2).

Until now, numerous adenosine receptor selective agonists and antagonists have been identified (see Table 7.1). The most selective agonists for A$_1$, A$_{2A}$ A$_{2B}$, and A$_3$ are CCPA, CGS21680, NECA, and Cl-IB-MECA, respectively. These ligands are closely related to adenosine in structure with only few modifications permitted. By using transgenic knockout mice, adenosine receptors have been demonstrated to regulate the central nervous system, immune system and cardiovascular function [19, 20]. In line with this finding, either A$_1$ or A$_3$ receptor overexpression protects against ischemic preconditioning in transgenic mice [21], emphasizing the leading role of purinoceptors for cellular and tissue functions.

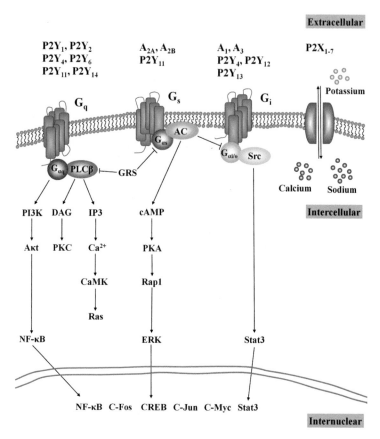

Figure 7.2 Signaling of purinergic receptors. The functional P2X receptors can mediate sodium and calcium influx and potassium efflux. The functional P1 and P2Y receptors belong to the G-coupled receptors (G_q, G_s and G_i). They signal trough PLC, cAMP and Stat3 pathways and finally modify the gene expression trough the transcription factors NF-kB, c-fos, CREB, c-jun, c-Myc and Stat3.

7.2.2 P2X Receptors

P2X receptors have been found to be expressed in almost all tissues and organs from the body, including nervous tissue, endothelium, smooth muscle, bone, epithelium and others [17, 23, 24]. They can assemble to form functional ion pores by homotrimers or heterotrimers [3, 25] (see Table 7.2). The functional P2X receptors can mediate sodium and calcium influx and potassium efflux (Figure 7.2), which leads to a membrane depolarization. The $P2X_7$ receptor has a unique subunit with a large intracellular tail, which con-

Table 7.2 Characteristics for P2X receptors.

	Potential co-assembly	Desensitization rate	Large organic cation permeability
$P2X_1$	X_1, X_2, X_3, X_5, X_6	rapid	unclear
$P2X_2$	X_1, X_2, X_3, X_5, X_6	very slow	NMDG
$P2X_3$	X_1, X_2, X_3, X_5	rapid	unclear
$P2X_4$	X_4, X_5, X_6	slow	NMDG
$P2X_5$	$X_1, X_2, X_3, X_4, X_5, X_6$	very slow	unclear
$P2X_6$	X_1, X_2, X_5, X_6	slow	unclear
$P2X_7$	X_7	slow	NMDG

tains a cell death domain regulating apoptosis via caspase 8/3 signalling [26]. It contains a putative SH3 binding domain controlling membrane blebbing through Rho [26], and a LPS binding domain modifying the transcription factors NF-κB, AP-1 and STAT6 via JAK2 and JNK signalling (Figure 7.2) [26, 27]. Compared to other tissues, the P2X receptors' presence and functional role is well documented in the nervous system. It has been proven to regulate fast synaptic transmission [28], astrocyte proliferation, neurotransmission and the generation of pain signals [29]. A variety of P2X receptor agonists and antagonists have been found or synthesized (see Table 7.1). However, some of P2X subtypes are still missing truly selective agonists or antagonists.

7.2.3 P2Y Receptors

P2Y receptors have been demonstrated to be widely distributed in several tissues, such as nervous tissue, endothelium, epithelium, muscle, bone, and immune cells [30]. As typical G-coupled proteins, they are composed of seven transmembrane-spanning domains and contain an intracellular C terminus, which can activate a set of intracellular signalling cascades (Figure 7.2). P2Y receptors can form heterotrimeric assemblies with adenosine receptors and may be capable to form homodimers as well [31]. $P2Y_{1,6,12,13}$ receptors are activated principally by nucleoside diphosphates, while other P2 receptors are triggered mainly by nucleoside triphosphates ($P2Y_{2,4,11}$). $P2Y_{2,4,6,11}$ receptors are activated by both purine and pyrimidine nucleotides, whereas $P2Y_{1,12,13}$ are stimulated by purines alone, and, uniquely, $P2Y_{14}$ is turned on by ribose-nucleotides [13, 32]. Almost all the P2Y receptors are coupled to G_q protein, except $P2Y_{12}$ and $P2Y_{13}$, which can activate a phosphoinositide-dependent PLC pathway with the following formation of inositol IP3 and intracellular Ca^{2+} mobilization. For instance, the $P2Y_2$ receptor, which is one of the best documented and a typical subtype of P2Y receptors, when crosslinked with integrin, can regulate NO expression by activating PLC, and mediate following Ca^{2+} and the PKC pathway (Figure 7.2). In addition, the $P2Y_2$ receptor can synergic cooperate with growth factor receptors, and thus regulate gene expression via ERK1/2, JNK and p38 MAPK pathways [17, 33]. Finally, the activation of the $P2Y_2$ receptor can cause proliferation or migration of human endothelial cells, smooth muscle cells and epithelial cells [17, 34–36]. Different from all other P2Y receptors, the $P2Y_{11}$ receptor has an unique ability to couple to both, G_q and G_s and thus consequently can mediate the PKA pathway via adenylate cyclase and Ca^{2+}, and the PKC pathway via PLC separately (Figure 7.2) [37]. To date, many agonists and antagonists have been utilized for P2 receptors, but some of them are not potent and others are not selective for the P2Y receptor subtypes (see Table 7.1).

7.3 Purinergic Receptors in Stem Cells

Stem cells which are mainly classified into embryonic stem cells (ESCs), adult stem cells, and recently also in induced pluripotent stem cells (iPS) have the ability of self-renewal and differentiation into various cell types. ESCs and iPS can differentiate into all kinds of cells, whereas adult stem cells have a limited plasticity [38, 39]. All stem cell types can be used as a key tool for the investigation of embryonic development and tissue renewal and the newly artificially developed iPS cells an be, in addition, an interesting new tool to investigate monogenetic disease in detail, since they can be produced from the patients itself. In addition, the unique ability of self-renewal and multipotent differentiation of stem cells provides new medical perspectives to rebuild cell and tissue functions and subsequently regenerate damage tissue or rebuild new organs in the future. The most prominent and long-term commonly established stem cells therapy is the haematopoietic stem cell transplantation from bone marrow to recreate healthy haematopoietic function after chemotherapy treating multiple myeloma or leukaemia [40, 41]. However, a complete and controllable differentiation is required for successful tissue reconstruction, which is still unachievable for most applications and stem cell types.

Since purinergic receptors take part in many important physiological and pathological processes, including cell proliferation, migration, secretion, and transmission these potent molecules can be expected to play a major role in stem cells and their signalling pathway towards differentiation as well. Although there is still a lot unclear as to how purinoceptors display their

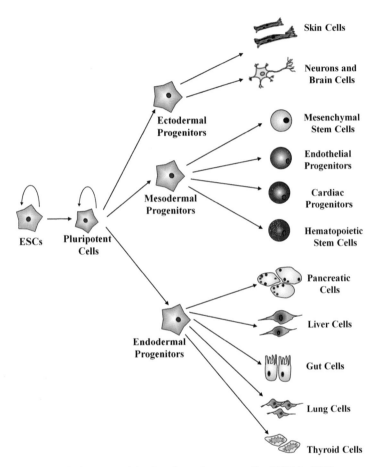

Figure 7.3 Differentiation potential of embryonic stem cells (ESCs). ESCs can generate tissue-specific stem cells and naturally have the ability to differentiate into all tissue specific cell lineages.

influence in the various stem cell types, there are already several publications indicating an important involvement of purinergic receptors in stem cell differentiation and proliferation, which will be reviewed in the following.

7.3.1 Purinergic Receptors in Embryonic Stem Cells

Embryonic stem cells (ESCs) generate tissue-specific stem cells and naturally have the ability to differentiate into all cell lineages (Figure 7.3) [42].

The role of purinoceptors has been investigated for this stem cell type itself and with respect to its differentiation potential mainly during the neuronal differentiation lineage. Adenosine-releasing embryonic stem cells have been generated by disruption of both alleles of adenosine kinase (AdK-/-) [43]. RT-PCR analysis showed $P2X_3$, $P2X_4$, $P2Y_1$, and $P2Y_2$ expression in mouse ESCs, and ATP stimulated cell proliferation through PKC, PI3K/Akt, and MAPKs pathways by activating P2 receptors [44]. Embryonic carcinoma cells derived from teratocarcinomas are regarded as the malignant counterparts of embryonic stem cells derived from the inner cell mass of blastocyst-stage embryos [45]. Thus, data achieved from these cells must be evaluated carefully. Nevertheless, they might present valuable insight in the signalling pathways in ESCs. Murine P19 embryonic carcinoma cells displayed a high expression of $P2X_3$, $P2X_4$, $P2Y_1$, and $P2Y_4$ receptors, which was decreased after induction of differentiation [46] whereas an increased gene and protein expression of $P2X_2$, $P2X_6$, $P2Y_2$, and $P2Y_6$ receptors was found during the process of differentiation in these cells. Resende and colleagues reported $P2X_{4/6}$ and $P2X_{2/6}$ heteromultimers to be present on P19 cells. They concluded that the $P2Y_1$ receptor was the major subtype involved in regulating cell proliferation and differentiation, followed by the $P2Y_2$ receptor and with a minor role for the $P2X_4$ receptor, results obtained by using different purinergic receptor agonists and antagonists [47]. These findings have been underlined by experiments using siRNA which showed that $P2X_4$ receptor down-regulation inhibited induction of proliferation of P19 cells [6].

P19 cell differentiation towards the neural lineage is severely influenced by long-term inhibition of $P2Y_1$ and possibly $P2X_2$ receptor activity, which causes a decreased activity of the NMDA receptor. The cholinergic receptor response can be inhibited by blockage of $P2Y_2$ and possibly $P2X_2$ receptors, which indicates that purinergic receptor activity is important for the definition of the neuronal phenotype [6]. GABAergic neurons derived from mouse ESCs can enhance $[Ca^{2+}]_i$ mainly by activating $P2X_2$, $P2X_4$ and $P2Y_1$ receptors, which indicates that mESCs may provide valuable models for investigating pharmacological neuronal function *in vitro* by using P2 signalling [48]. Taken together, purinoceptors play an important role in embryonic stem cell differentiation towards the neuronal lineage but little is know today about their role in the differentiation processes of ESCs towards other lineages.

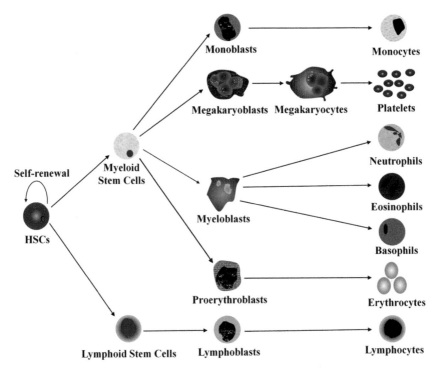

Figure 7.4 Differentiation potential of hematopoietic stem cells (HSCs). HSCs belong to the multipotent stem cells. They can give rise to all types of blood cells via some precursors (monoblasts, megakaryoblasts, myeloblasta, proerythroblasts and lymphoblasts) differentiating into monocytes, platelets, neutrophils, eosinophils, basophils, erythrocytes, and lymphocytes which are composed out of T-cells, B-cells, and NK-cells (not shown).

7.3.2 Purinergic Receptors in Hematopoietic Stem Cells

Hematopoietic stem cells (HSCs) are multipotent adult stem cells and can give rise to all types of blood cells, such as monocytes, macrophages, neutrophils, platelets, dendritic cells, T-cells, B-cells, NK-cells and others (Figure 7.4) [49].

Nucleotides have been proven to maintain the presence of some specific growth factors (e.g. IL-3), which are indispensable for HSCs' survival, proliferation and development [50]. This is in line with the finding that the adenosine A_3 receptor enables the hematopoietic growth factors and some cytokines to take effect, which influenced the growth of hematopoietic progenitor cells for granulocytes and macrophages in mouse bone marrow cells *in vitro*, data achieved by using its specific agonist IB-MECA [51]. Furthermore, Hofer and his colleagues demonstrated a significant increase of proliferation of mouse HSCs, both, in the resting state and in the depletion phase, but there was no detectable proliferation in the regeneration phase by stimulating with IB-MECA [52]. CPA as a selective agonist of A_1 receptor displayed an opposite effect, which both indicates the plasticity and homeostatic role of the adenosine receptor expression [53]. The administration AMP or the drug dipyridamole can induce Adenosine receptor signalling via the elevation of extracellular adenosine. This can enhance the cycling of the hematopoietic progenitor cells and hematopoiesis in mice, which might have pharmacological implications in the therapy of blood disorders in humans, if applicable there too [54]. There are already data for the application in humans for another P1 receptor. Adenosine A_{2A} receptor activation limits graft-versus-host disease after allogenic HSCs [55].

Kalambakas and co-workers claimed that some ADP specific receptors were present on both human and rat hematopoietic cell lines [56], which probably indicates that the ADP purinoceptors such as $P2Y_1$, $P2Y_{12}$, and $P2Y_{13}$ can play a role. ATP can be stored in vesicles and released in a calcium-sensitive manner in hematopoietic stem cells. P2X receptor expression was observed in HSCs when induced pharmacologically by using specific antagonists, leading to hematopoietic progenitor proliferation, but not to myeloid differentiation *in vitro*. This data was underlined by *in vivo* studies in mice suffering from chronic inflammation, where HSCs were expanded remarkable and their cycling activity was sensitive to the P2X receptor antagonist oxide ATP [57]. ATP and UTP were found to induce human $CD34^+$ HSCs proliferation both, *in vitro* and *in vivo*, with the detected presence of all P2X subtypes, and $P2Y_1$ and $P2Y_2$. Generally $P2X_1$ and $P2X_7$ were found in all kinds of human hematopoietic cells, but there with no detection of $P2X_4$ and $P2X_5$. Additionally, $P2X_1$ and $P2X_7$ could be down-regulated after ATP stimulation in HSCs [58]. However, stimulation with high concentrations of ATP led to a remarkable inhibition of cell proliferation despite the presence of cytokine stimulation. In addition, Yoon and his colleagues found that only the $P2X_1$ and $P2X_7$ receptor were present in murine hematopoietic cells without ATP induction and that the $P2X_7$ receptor was expressed more strongly in murine hematopoietic progenitor cells, if compared to that in HSCs. Both suramin and oxidized ATP, which are unselective antagonists for P2X receptors, can inhibit the induced cell death of hematopoietic cells [59]. This data can have a direct impact on the clinical use of hematopoietic stem cells. For example, the $P2X_7$ receptor polymorphism can influence the clinical outcome in HLA-matched sibling allogeneic hematopoietic stem cell transplantation via secretion of cytokines such as IL-1 [60].

We will now look at P2Y receptors signalling in bone-marrow HSCs. The $P2Y_{14}$ receptor was firstly found as a chemoattractant protein which may be involved in cell localization and migration of HSCs [61]. Functional $P2Y_{11}$ receptors seemed to be involved in the process of maturating monocyte-derived dendritic cells and granulocytic differentiation into the promyelocyte lineages [62]. Extracellular UTP improved HSCs migration by down-regulating CXCR4 and increasing cell adhesion fibonectin remarkably in CXCL12 simulated human $CD34^+$ HSCs *in vitro*. On the other hand, *in vivo* data using immunodeficient (NOD/SCID) mice showed that a preincubation with UTP improved the bone marrow homing efficiency of human $CD34^+$ HSCs remarkably [63], which suggests that UTP purinoceptors such as $P2Y_2$, $P2Y_4$, $P2Y_{14}$ may have a crucial role in HSCs.

Taken together, there is already quite a lot of information on purinoceptor signalling in HSCs, the hematopoietic progenitor and the consecutive following differentiated hematopoietic cells as well as in their impact on stem cell transplantation. This is very different for mesenchymal stem cells.

7.3.3 Purinergic Receptors in Mesenchymal Stem Cells

Mesenchymal stem cells (MSCs) are another multipotent type of adult stem cells. They can be isolated from a variety of mesodermal tissues including dental pulp, skin, tendon, peripheral blood, umbilical cord blood, bone marrow stroma and adipose tissue, the last three being the most prominent sources. MSCs have the potential to differentiate into multiple mesenchymal lineages, such as osteogenic, adipogenic, chondrogenic cells and others (Figure 7.5) [64, 65].

Adenosine receptors have been demonstrated to regulate osteogenesis versus adipogenesis in MSCs derived from bone marrow [66]. Human bone marrow stromal cells not only express all adenosine receptors, but also secrete adenosine and CD73. Adenosine on the other hand has a potent stimulatory effect on IL-6 secretion, but inhibits osteoprotegerin secretion. This suggests that adenosine receptors are key molecules in regulating the osteoprogenitor cell differentiation and therefore may mediate bone formation and resorption [67]. Additionally, adenosine can also trigger some key endodermal and hepatocyte-specific gene expression in mouse and human MSCs in vitro [68].

Endogenously produced adenosine also plays a significant role in promoting mouse bone marrow MSC proliferation and differentiation through the

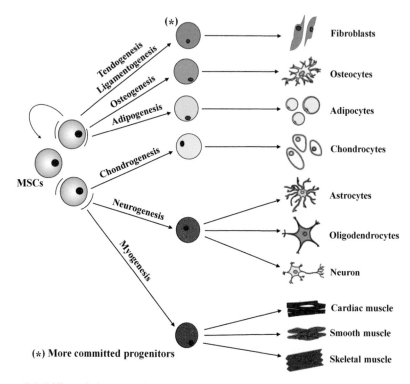

Figure 7.5 Differentiation potential of mesenchymal stem cells (MSCs). MSCs belonged to adult stem cells. In accordance with the hematopoietic stem cells more committed progenitor cells are discussed, but they have not been defined yet. MSCs have the potential to differentiate into multiple mesenchymal derived lineages. After tendogenesis and ligamentogenesis e.g. fibroblasts are produced. Osteogenesis leads to osteocytes, and adipogenesis to adipocytes. Similarly, chondrogenesis will end with chondrodrocytes, and neurogenesis will lead to astrocytes, oligodendrocytes, and neurons. Myogenesis will generate cardiac, smooth, and skeletal muscle cells.

adenosine A_{2A} receptor [69]. Using agonists of the A_{2A} receptor, it was shown that A_{2A} also promotes wound healing and tissue regeneration by enhancing the accumulation of bone marrow derived endothelial progenitor cells [70]. The A_{2A} receptor was later shown to significantly inhibit MSCs chemotaxis in response to the chemoattractant HGF in both, human and mouse bone marrow derived MSCs. This is mediated by the up-regulation of cAMP and the PKA pathway which can inhibit cytosolic calcium signaling, down-regulate HGF-induced Rac1, and finally suppress HGF-induced stress fiber formation.

The up-regulation of extracellular adenosine or dipyridamole has been demonstrated to protect bone marrow from radiation and chemotherapy injury. These molecules can even improve responses following radiation or chemotherapy via adenosine A_1 and A_3 receptors [71, 72] demonstrating a direct possible use of P1 signalling in MSCs.

P2 receptors were first described by Wang and his colleagues to be expressed in $CD34^+$ stem and progenitor cells from bone marrow. They demonstrated a relative highly expression for the types $P2X_1$, $P2X_4$, $P2Y_1$, and $P2Y_2$ [73]. Riddle's group verified the widespread expression of P2 receptors in bone marrow MSCs, with a high presence of $P2Y_2$ and $P2X_7$, which have an implication in bone cell mechanotransduction [74]. $P2X_7$ mice knockout experiments displayed a distinct skeletal phenotype, including a decreased periosteal bone formation rate and an increased trabecular bone resorption, resulting from a decreased sensitivity to mechanical signals [75, 76].

Several data have been achieved by using the agonist ATP. Oscillatory fluid flow induces the vesicular release of ATP from human bone marrow MSCs, which directly mediates bone marrow MSCs proliferation [77]. This is in contrast to data showing that ATP released from early passages (P0–P5) of cultivating human bone marrow MSCs decreases cell proliferation, and that the unselective P2 antagonist PPADS and the selective $P2Y_1$ antagonist 2'-deoxy-N6-methyladenosine 3', 5'-bisphosphate (MRS 2179) could induce human bone marrow MSCs proliferation. This might be explained if considering the effect of P2Y together with P2X receptor effects. Outward ATP-sensitive currents were activated through P2Y receptors, otherwise inward ATP-sensitive currents were evoked by activating P2X receptors [78]. ATP and UTP have a stronger potency to evoke intracellular Ca^{2+} levels than ADP and UDP in bone marrow stromal cells [79]. Kawano and his co-workers demonstrated that the $P2Y_1$ receptor can activate PLC-β, produce IP3, and induce spontaneous $[Ca^{2+}]_i$ oscillations in human bone marrow MSCs, when they are activated by autocrine/paracrine secreted ATP [80]. In addition the presence of the $P2Y_2$ receptor was confirmed by both, RT-PCR and immunohistochemical staining. Suppression by U73122 and suramin was also observed, but no suppression by PPADS has been shown [79].

Human adipose tissue derived mesenchymal stem cells (ATSCs), which can be easily isolated from liposuction material, a by-product in plastic surgeries, can be maintained and expanded *in vitro* for long time periods. They can differentiate into chondrogenic, adipogenic, osteogenic and myogenic lineages [81, 82]. In addition, several publications also claim lineage differentiations achieved from these cells which are naturally derived from the endoderm [83, 84]. Thirteen out of fifteen P2 receptors were claimed to be expressed in human ATSCs, with the only absence of $P2X_1$ and $P2X_2$. Interestingly, down-regulation of $P2Y_4$ and $P2Y_{14}$ were found after 4 weeks adipogenic differentiation of ATSCs, and osteogenic differentiation of ATSCs in both gene and protein level. An up-regulation of $P2X_6$ in adipogenic differentiation, but a down-regulation in osteogenic differentiation suggests these receptors to be involved in ATSCs differentiation [85]. Both ATP and UTP can induce Ca^{2+} influx suggesting P2 receptor activity in ATSCs, while suramin and PPADS can inhibit osteogenic differentiation induced by ATP but not by UTP [86] pin-pointing presumably to $P2Y_1$ $P2Y_4$ and $P2Y_6$. The $P2X_5$, $P2X_7$, $P2Y_1$ and $P2Y_2$ receptors may partake in osteogenic differentiation, whereas $P2Y_{11}$ up-regulation induced lipolysis and leptin during the differentiation towards adipocytes [87].

Remarkable MSCs also seem to be endocrinally and/or paracrinally active. The up-regulation of IL-6, VEGF-A, HGF, and TGF-β were found in human MSCs derived from umbilical cord and human MSCs isolated from bone marrow by ELISA. However, only human pancreatic islets co-cultured with umbilical cord MSCs showed an increased insulin secretory function and ATP content [88].

Few data are for MSCs derived from new sources such as skin. Skin tissue is easy accessibly with non-invasive methods. It could be shown that extracellular ATP also mediates an intracellular calcium increase through ionotropic P2X receptors in human MSCs derived from human skin tissues [89].

Taken together purinoceptors also play a crucial role in mesenchymal stem cell proliferation and differentiation. But more studies are necessary to fully discover their function and thus enable their use in clinical approaches.

7.4 Conclusions

Much interest and attention has been paid to stem cells with their special characteristic of self-renewal and differentiation potential, which pinpoints to them as good future sources for Regenerative Medicine [90]. However, how to induce stem cell differentiation toward the desired direction without having stem cells not reacting to the stimulus within is unclear. Also, how to trigger a differentiation lineage until the end, to achieve a fully differentiated cell type is an unsolved problem. Both might lead to tumorigenesis, if applied in humans. Basic research must be increased to better understand the signalling pathways leading to the above mentioned goals.

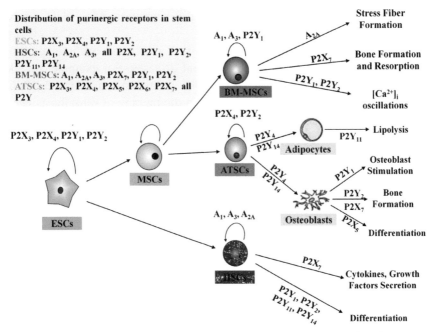

Figure 7.6 Distributions of purinergic receptors in various stem cells. Purinergic receptors (P1, P2X, P2Y receptors) are expressed in different stem and progenitor cells each in a unique pattern of combinations. They can be expected to play an important role in determining stem cell fate by regulating cell proliferation and differentiation.

P1, P2X and P2Y receptors have not only been observed to be expressed in most stem and progenitor cells, but also to play a functional role in their proliferation, migration and differentiation (see Figure 7.6 for a summary).

However, their down-stream signal pathways and their natural regulatory mechanism are still unclear to some extent. Notably, the same purinergic receptor subtype may induce different, sometimes even opposite effect on various stem cells. For example, as described above, the P2Y$_1$ receptor has been shown to induce proliferation in embryonic stem cells [44] and in P19 embryonic carcinoma cells [46]. In contrast to this P2Y$_1$ inhibits proliferation in the early stages of human bone marrow derived MSCs [80], suggesting a time and cell type specific effect. Specific agonists and antagonists for some purinergic receptors are still missing, which impedes research to understand purinergic signalling and its activation mechanism in stem cells. Recently, using small interference RNA or gene knock-out animal models to investigate individual purinergic receptor subtype *in vitro* and *in vivo* led to new data for purinergic receptor function in stem cells differentiation. More data can be expected by these approaches.

There are already several drugs approved, which are artificial purinoceptor ligands to treat various very different diseases. For example, for the P2Y receptor linked to regulation of platelet aggregation and activation, clopidogrel an irreversible inhibitor, was used in the therapy of thrombosis [91]. It can be expected that other artificial purinergic receptor ligands could be used to trigger stem cell fate in the future and thus make an important additional step towards the safe use of stem cell derived regenerative medicine approaches.

7.5 List of Abbreviations

ADP	adenosine 5'-diphosphate
ADPβS	adenosine 5'-O-(2-thiodiphosphate)
ALT-146e	4-{3-[6-Amino-9-(5-ethylcarbamoyl-3,4-dihydroxy-tetrahydro-furan-2-yl)-9H-purin-2-yl]-prop-2-ynyl}-cyclohexanecarboxylic acid methyl ester
AMP	adenosine 5'-monophosphate
AMPαS	adenosine 5'-O-(thiomonophosphate)
Ap$_4$A	diadenosine tetraphosphate
Ap$_5$A	diadenosine pentaphosphate
ATP	adenosine 5'-triphosphate
ATPγS	adenosine 5'-O-(3-thiotriphosphate)
ATSCs	human adipose tissue derived mesenchymal stem cells
A3P5PS	adenosine 3'-phosphate 5'-phosphosulphate
BBG	brilliant blue green
BzATP	2'-3'-O-(4-benzoyl-benzoyl)-ATP
cAMP	cyclic AMP
CCPA	chlorocyclopentyl adenosine
CD73	ecto-5'-nucleotidase
CHA	N^6-Cyclohexyladenosine
CPA	N^6-cyclopentyl adenosine
CPT	8-Cyclopentyl-1,3-dimethylxanthine
CPX	8-Cyclopentyl-1,3-dipropylxanthine
CTP	cytosine triphosphate
CTV510	N-(3(R)-Tetrahydrofuranyl)-6-aminopurine riboside
DFCs	follicle precursor cells
DPMA	N^6-[2-(3,5-Dimethoxyphenyl)-2-(2-methylphenyl)ethyl]

	adenosine
ESCs	embryonic stem cells
GTP	guanosine 5'-triphosphate
HGF	hepatocyte growth factor
HSCs	hematopoietic stem cells
IB-MECA	N^6-(3-iodobenzyl) adenosine-5'-N methyluronamide
IDP	Inosine 5'-diphosphate
IL-1	interleukin-1
IL-6	interleukin-6
INS37217	P(1)-(Uridine 5')-P(4)- (2'-deoxycytidine 5') tetraphosphate tetrasodium salt
INS365	diuridine tetraphosphate
IP3	inosine triphosphate
Ip5I	di-inosine pentaphosphate
iPS	induced pluripotent stem cells
2-MeSADP	2-methylthio ADP
2-MeSATP	2-methylthio ATP
2-MeSAMP	2-Methylthioadenosine-5'-monophosphate
α,β-MeATP	α,β-Methylene-adenosine-5'-triphosphate
MeSADP	methylthio ADP
MeSATP	methylthio ATP
MRS1754	8-[4-[[(4-Cyano)phenylcarbamoylmethyl]oxy]phenyl]-1,3-di-(n-propyl)xanthine
MRS 2179	2'-Deoxy-N^6-methyl adenosine 3', 5'-bisphosphate
MRS 2279	(N)-Methanocarba-N^6-methyl-2-chloro-2'-deoxyadenosine-3',5'-bisphosphate
MRS2365	(N)-Methanocarba-2-Methylthioadenosine-5'-diphosphate
MRS2578	1,4-di-(Phenylthioureido) butane
MRS2500	(N)-Methanocarba-N^6-methyl-2-iodo-2'-deoxyadenosine-3',5'-bisphosphate
MSCs	mesenchymal stem cells
N-0840	N^6-Cyclopentyl-9-methyladenine
NECA	5-N-ethylcarboxamido adenosine
NMDG	D-methyl-D-Glucamine
NO	nitric oxide
NTP-ATP	2',3'-O-(2,4,6-Trinitrophenyl) adenosine triphosphate
RB2	reactive blue 2
PGE2	prostaglandin E2
PKC	protein kinase C

PLC	phospholipase C
PPADS	pyridoxal-5-phosphate-6-azophenyl-2',4'-disulphonic acid
PPNDS	pyridoxal-5-phosphate-6-(2'-naphthylazo-6-nitro-4',8'-disulphonate)
siRNA	small interference RNA
TGF-β	transforming growth factor-β
TNP-ATP	2',3'-O-(2,4,6-Trinitrophenyl) adenosine triphosphate
UDP	uridine-5'-diphosphate
UTP	uridine-5'-triphosphate
UTPγS	uridine 5'-O-(3-thiotriphosphate)
VEGF-A	vascular endothelial growth factor-A
WRC-0571	8-(N-Methylisopropyl)amino-N-(5'-endohydroxy-endonorbornyl)-9-methyladenine

Acknowledgements

This work was supported by China Scholarship Council No. 20100602024; German Aerospace Center (DLR) for YZ and BMBF-AIF, AdiPaD; FKZ: 1720X06 for ET.

References

[1] Burnstock, G., Knight, G.E, Cellular distribution and functions of P2 receptor subtypes in different systems, *International Review Cytology*, 240:301–304, 2004.

[2] Burnstock, G., Purinergic nerves, *Pharmacological Review*, 24(3):509–581, September 1972.

[3] Burnstock, G., Purine and pyrimidine receptors, *Cellular and Molecular Life Sciences*, 64(12):1471–1483, June 2007.

[4] Burnstock, G., Williams, M., P2 purinergic receptors: Modulation of cell function and therapeutic potential, *Journal of Pharmacology and Experimental Therapy*, 295(3):862–869, December 2000.

[5] Burnstock, G., Unresolved issues and controversies in purinergic signalling, *Journal of Physiology*, 586(14):3307–3312, July 2008.

[6] Burnstock, G., Ulrich, H., Purinergic signaling in embryonic and stem cell development, *Cellular Molecular Life Science*, 68(8):1369–1394, April 2011.

[7] Dranoff, J.A., O'Neill, A.F., Franco, A.M., Cai, S.Y., Connolly, G.C., Ballatori, N., A primitive ATP receptor from the little skate Raja erinacea, *Journal of Biological Chemistry*, 275(39):30701–30706, September 2000.

[8] Burnstock, G., A basis for distinguishing two types of purinergic receptor. In: *Cell Membrane Receptors for Drugs and Hormones: A Multidisciplinary Approach*, Straub, R.W. and Bolis, L. (eds.), Raven Press, New York, pp. 107–118, 1978.

[9] Olah, M.E., Stiles, G.L., The role of receptor structure in determining adenosine receptor activity, *Pharmacology & Therapeutics*, 85(2):55–75, February 2000.

[10] St Hilaire, C., Carroll, S.H., Chen, H., Ravid, K., Mechanisms of induction of adenosine receptor genes and its functional significance, *Journal of Cellular Physiology*, 218(1):35–44, January 2009.

[11] Cobb, B.R. and Clancy, J.P., Molecular and cell biology of adenosine receptors, *Current Topic Membrane*, 54:151–181, 2003.

[12] Buell, G., Collo, G., Rassendren, F., P2X receptors: An emerging channel family, *European Journal of Neurobiology*, 8(10):2221–2228, October 1996.

[13] Surprenant, A., North, R.A., Signaling at purinergic P2X receptors, *Annual Review of Physiology*, 71:333–359, 2009.

[14] von Kügelgen, I., Wetter, A., Molecular pharmacology of P2Y-receptors, *Naunyn–Schmiedeberg's Archives Pharmacology*, 362(4/5), 310–323, 2000.

[15] Jacobson, K.A., Jarvis, M.F., Williams, M., Purine and pyrimidine (P2) receptors as drug targets, *Journal of Medicinal Chemistry*, 45(19):4057–4093, September 2002.

[16] Abbracchio, M.P., Burnstock, G., Boeynaems, J.-M., Barnard, E.A., Boyer, J.L., Kennedy, C., Knight, G.E., Fumagalli, M., Gachet, C., Jacobson, K.A., Weisman, G.A., International Union of Pharmacology LVIII: Update on the P2Y G protein-coupled nucleotide receptors: From molecular mechanisms and pathophysiology to therapy, *Pharmacological Review*, 58:281–341, 2006.

[17] Erb, L., Liao, Z., Seye, C., Weisman, G., P2 receptors: Intracellular signalling, *Pflugers Arch-European Journal of Physiology*, 452(5):552–562, August 2006.

[18] Reshkin, S.J., Guerra, L., Bagorda, A., Debellis, L., Cardone, R., Li, A.H., Jacobson, K.A. and Casavola, V., Activation of A_3 adenosine receptor induces calcium entry and chloride secretion in A(6) cells, *Journal of Membrane Biology*, 178(2):103–113, November 2000.

[19] Sun, D., Samuelson, L.C., Yang, T., Huang, Y., Paliege, A., Saunders, T., Briggs, J. and Schnermann, J., Mediation of tubuloglomerular feedback by adenosine: Evidence from mice lacking adenosine 1 receptors, *Proceedings of National Academy Science of United States of America*, 98(17):9983–9988, August 2001.

[20] Ledent, C., Vaugeois, J.M., Schiffmann, S.N., Pedrazzini, T., El Yacoubi, M., Vanderhaeghen, J.J., Costentin, J., Heath, J.K., Vassart, G. and Parmentier, M., Aggressiveness, hypoalgesia and high blood pressure in mice lacking the adenosine A_{2A} receptor, *Nature*, 388(6643):674–678, August 1997.

[21] Lankford, A.R., Yang, J.N., Rose Meyer, R., French, B.A., Matherne, G.P., Fredholm, B.B. and Yang, Z., Effect of modulating cardiac A1 adenosine receptor expression on protection with ischemic preconditioning, *American Journal of Physiology-Heart and Circulatory Physiology*, 290(4):1469–1473, April 2006.

[22] Burnstock, G., Pathophysiology and therapeutic potential of purinergic signaling, *Pharmacological Review*, 58(1):58–86, March 2006.

[23] North, R.A., Molecular physiology of P2X receptors, *Physiological Review*, 82(4):1013–1067, October 2002.

[24] Vial, C., Roberts, J.A., Evans, R.J., Molecular properties of ATP-gated P2X receptor ion channels, *Trends in Pharmacological Science*, 25(9):487–493, September 2004.

[25] Nicke, A., Baumert, H.G., Rettinger, J., Eichele, A., Lambrecht, G., Mutschler, E. and Schmalzing, G., P2X$_1$ and P2X$_3$ receptors form stable trimers: A novel structural motif of ligand-gated ion channels, *EMBO Journal*, 17(11):3016–3028, June 1998.

[26] Denlinger, L.C., Fisette, P.L., Sommer, J.A., Watters, J.J., Prabhu, U., Dubyak, G.R., Proctor, R.A. and Bertics, R.J., Cutting edge: the nucleotide receptor P2X$_7$ contains multiple protein- and lipid-interaction motifs including a potential binding site for bacterial lipopolysaccharide, *Journal of Immunology*, 167(4):1871–1876, August 2001.

[27] Denlinger L.C., Sommer, J.A., Parker, K., Gudipaty, L., Fisette, P.L., Watters, J.W., Proctor, R.A., Dubyak, G.R. and Bertics, R.J., Mutation of a dibasic amino acid motif within the C terminus of the P2X$_7$ nucleotide receptor results in trafficking defects and impaired function, *Journal of Immunology*, 171(3):1304–1311, August 2003.

[28] Abbracchio, M.P., Burnstock, G., Purinoceptors: Are there families of P2X and P2Y purinoceptors? *Pharmacological Therapy*, 64(3):445–475, 1994.

[29] Burnstock, G., Purinergic signalling, *British Journal of Pharmacology*, 147(Suppl 1):172–181, January 2006.

[30] Neary, J.T., Kang, Y., Bu, Y., Yu, E., Akong, K., Peters, C.M., Mitogenic signaling by ATP/P2Y purinergic receptors in astrocytes: Involvement of a calcium-independent protein kinase C, extracellular signal-regulated protein kinase pathway distinct from the phosphatidylinositol-specific phospholipase C/calcium pathway, *Journal of Neuroscience*, 19(11):4211–4220, June 1999.

[31] Yoshioka, K., Hosoda, R., Kuroda, Y. and Nakata, H., Hetero-oligomerization of adenosine A1 receptors with P2Y1 receptors in rat brains, *FEBS Letters*, 531(2):299–303, November 2002.

[32] King, B.F., Townsend-Nichlson, A., Burnstock, G., Metabotropic receptors for ATP and UTP: Exploring the correspondence between native and recombinant nucleotide receptors, *Trends in Pharmacological Science*, 19(12):506–514, December 1998.

[33] Seye, C.I., Kong, Q., Erb, L., Garrad, R.C., Krugh, B., Wang, M., Turner, J.T., Srurek, M., Gonzalez, F.A. and Weisman, G.A., Functional P2Y$_2$ nucleotide receptors mediate uridine 5-triphosphate-induced intimal hyperplasia in collared rabbit carotid arteries, *Circulation*, 106(21):2720–2726, November 2002.

[34] Wilden, P.A., Agazie, Y.M., Kaufman, R., Halenda, S.P., ATP stimulated smooth muscle cell proliferation requires independent ERK and PI3K signaling pathways, *American Journal of Physiology*, 275(4, pt 2):1209–1215, October 1998.

[35] Ahn, J.S., Camden, J.M., Schrader, A.M., Redman, R.S., Turner, J.T., Reversible regulation of P2Y(2) nucleotide receptor expression in the duct-ligated rat submandibular gland, *American Journal of Physiology-Cell Physiology*, 279(2):286–294, August 2000.

[36] Seye, C.I., Yu, N., Jain, R., Kong, Q., Minor, T., Newton, J., Erb, L., Gonzalez, F.A. and Weisman, G.A., The P2Y$_2$ nucleotide receptor mediates UTP-induced vascular cell adhesion molecule-1 expression in coronary artery endothelial cells, *Journal of Biological Chemistry*, 278(27):24960–24965, July 2003.

[37] Qi, A.D., Kennedy, C., Harden, T.K., Nicholas, R.A., Differential coupling of the human P2Y (11) receptor to phospholipase C and adenylyl cyclase, *British Journal of Pharmacology*, 132(1):318–326, January 2001.

[38] Donovan, P.J., Gearhart, J., The end of the beginning for pluripotent stem cells, *Nature*, 414(6859):92–97, November 2001.

[39] Liu, S., iPS cells: A more critical review, *Stem Cells and Development*, 17(3):391–397, June 2008.

[40] Bladé, J., Samson, D., Reece, D., Criteria for evaluating disease response and progression in patients with multiple myeloma treated by high-dose therapy and haemopoietic stem cell transplantation. Myeloma Subcommittee of the EBMT. European Group for Blood and Marrow Transplant, *British Journal of Haematology*, 102(5):1115–1123, September 1998.

[41] Pavletic, S., Khouri, I., Haagenson, M., Unrelated donor marrow transplantation for B-cell chronic lymphocytic leukemia after using myeloablative conditioning: results from the Center for International Blood and Marrow Transplant research, *Journal of Clinical Oncology*, 23(24):5788–5794, August 2005.

[42] Martin, G.R., Isolation of a pluripotent cell line from early mouse embryos cultured in medium conditioned by teratocarcinoma stem cells, *Proceedings of National Academy Science of United States of America*, 78(12): 7634–7638, December 1981.

[43] Fedele, D.E., Koch. P., Scheurer, L., Simpson, E.M., Möhler, H., Brüstle, O., Boison, D., Engineering embryonic stem cell derived glia for adenosine delivery, *Neuroscience Letters*, 370(2/3): 160–165, November 2004.

[44] Heo, J.S., Han, H.J., ATP stimulates mouse embryonic stem cell proliferation via protein kinase C, phosphatidylinositol 3-kinase/Akt, and mitogen-activated protein kinase signaling pathways, *Stem Cells*, 24(12):2637–2648, December 2006.

[45] Andrews, P.W., Matin, M.M., Bahrami, A.R., Damjanov, I., Gokhale, P. Draper, J.S., Embryonic stem (ES) cells and embryonal carcinoma (EC) cells: Opposite sides of the same coin, *Biochemical Society Transactions*, 33(Pt 6):1526–1530, December 2005.

[46] Resende, R.R., Majumder, P., Gomes, K.N., Britto, L.R. Ulrich, H., P19 embryonal carcinoma cells as *in vitro* model for studying purinergic receptor expression and modulation of N-methyl-D-aspartate-glutamate and acetylcholine receptors during neuronal differentiation, *Neuroscience*, 146(3):1169–1181, May 2007.

[47] Resende, R.R., Britto, L.R., Ulrich, H., Pharmacological properties of purinergic receptors and their effects on proliferation and induction of neuronal differentiation of P19 embryonal carcinoma cells *International Journal of Developmental Neuroscience*, 26(7):763–777, November 2008.

[48] Khaira, S.K., Pouton, C.W., Haynes, J.M., $P2X_2$, $P2X_4$ and $P2Y_1$ receptors elevate intracellular Ca^{2+} in mouse embryonic stem cell-derived GABAergic neurons, *British Journal of Pharmacology*, 158(8):1922–1931, December 2009.

[49] Muller-Sieburg, C.E., Cho, R.H., Thoman, M., Adkins, B., Sieburg, H.B., Deterministc regulation of hematopoietic stem cell self-renewal and differentiation, *Blood*, 100(4):1302–1309, August 2002.

[50] Whetton, A.D., Huang, S.J., Monk, P.N., Adenosine triphosphate can maintain multipotent haemopoietic stem cells in the absence of interleukin 3 via a membrane permeabilization mechanism, *Biochemical and Biophysical Research Communications*, 152(3):1173–1178, May 1988.

[51] Hofer, M., Vacek, A., Pospisil, M., Hola, J., Streitova, D., Znojil, V., Activation of adenosine A3 receptors potentiates stimulatory effects of IL-3, SCF, and GM-CSF on mouse granulocytemacrophage hematopoietic progenitor cells, *Physiological Research*, 58(2):247–252, 2009.

[52] Hofer, M., Vacek, A., Pospisil, M., Weiterova, L., Hola, J., Streitova, D., Znojil, V., Adenosine potentiates stimulatory effects on granulocyte-macrophage hematopoietic progenitor cells *in vitro* of IL-3 and SCF, but not those of G-CSF, GM-CSF and IL-11, *Physiological Research*, 55(5):591–596, 2006.

[53] Hofer, M., Pospisil, M., Znojil, V., Holá, J., Streitová, D., Vacek, A., Homeostatic action of adenosine A_3 and A_1 receptor agonists on proliferation of hematopoietic precursor cells, *Experimental Biology and Medicine*, 233(7):897–900, June 2008.

[54] Pospíšil, M., Hofer, M., Vacek, A., Netíková, J., Holá, J., Znojil, V., Weiterová, L., Drugs elevating extracellular adenosine enhance cell cycling of hematopoietic progenitor cells as inferred from the cytotoxic effects of 5-fluorouracil, *Experimental Hematology*, 29(5):557–562, May 2001.

[55] Lappas, C.M., Liu, P.C., Linden, J., Kang, E.M., Malech, H.L., Adenosine A_{2A} receptor activation limits graft-versus-host disease after allogenic hematopoietic stem cell transplantation, *Journal of Leukocytes Biology*, 87(2):345–354, February 2010.

[56] Kalambakas, S.A., Robertson, F.M., O'Connell, S.M., Sinha, S., Vishnupad, K., Karp, G.I., Adenosine diphosphate stimulation of cultured hematopoietic cell lines, *Blood*, 81(10):2652–2657, May 1993.

[57] Casati, A., Frascoli, M., Traggiai, E., Proietti, M., Schenk, U., Grassi, F., Cell-autonomous regulation of hematopoietic stem cell cycling activity by ATP, *Cell Death and Differentiation*, 18(3):396–404, March 2011.

[58] Lemoli, R.M., Ferrari, D., Fogli, M., Rossi, L., Pizzirani, C., Forchap, S., Chiozzi, P., Vaselli, D., Bertolini, F., Foutz, T., Aluigi, M., Baccarani, M., Di Virgilio, F., Extracellular nucleotides are potent stimulators of human hematopoietic stem cells *in vitro* and *in vivo*, *Blood*, 104(6):1662–1670, September 2004.

[59] Yoon, M.J., Lee, H.J., Lee, Y.S., Kim, J.H., Park, J.K., Chang, W.K., Shin, H.C. and Kim, D.K., Extracellular ATP is involved in the induction of apoptosis in murine hematopoietic cells, *Biological and Pharmaceutical Bulletin*, 30(4):671–676, April 2007.

[60] Lee, K.H., Park, S.S., Kim, I., Kim, J.H., Ra E.K., Yoon, S.S., Hong, Y.C., Park, S. and Kim, B.K., P2X7 receptor polymorphism and clinical outcomes in HLA-matched sibling allogeneic hematopoietic stem cell transplantation, *Haematologica*, 92(5):651–657, May 2007.

[61] Lee, B.C., Cheng, T., Adams, G.B., Attar, E.C., Miura, N., Lee, S.B., Saito, Y., Olszak, I., Dombkowski, D., Olson, D.P., Hancock, J., Choi, P.S., Haber, D.A., Luster, A.D. and Scadden, D.T., P2Y-like receptor, GPR105 (P2Y$_{14}$), identifies and mediates chemotaxis of bone-marrow hematopoietic stem cells, *Genes and Development*, 17(13):1592–1604, July 2003.

[62] Wilkin, F., Duhant, X., Bruyns, C., Suarez-Huerta, N., Boeynaems, J.M., Robaye, B., The P2Y$_{11}$ receptor mediates the ATP-induced maturation of human monocyte-derived dendritic cells, *Journal of Immunology*, 166(12):7172–7177, June 2001.

[63] Rossi, L., Manfredini, R., Bertolini, F., Ferrari, D., Fogli, M., Zini, R., Salati, S., Salvestrini, V., Gulinelli, S., Adinolfi, E., Ferrari, S., Di Virgilio, F., Baccarani, M. and Lemoli, R.M., The extracellular nucleotide UTP is a potent inducer of hematopoietic stem cell migration, *Blood*, 109(2):533–542, January 2007.

[64] Pittenger, M., Mackay, A., Beck, S., Jaiswal, R.K., Douglas, R., Mosca, J.D., Moorman, M.A., Simonetti, D.W., Craig, S. and Marshak, D.R., Multilineage potential of adult human mesenchymal stem cells, *Science*, 284(5411):143–147, April 1999.

[65] Phinney, D.G., Prockop, D.J., Concise review: Mesenchymal stem/multipotent stromal cells: The state of transdifferentiation and modes of tissue repair-current views, *Stem Cell*, 25(11):2896–2902, November 2007.

[66] Gharibi, B., Elford, C., Lewis, B.M., Ham, J., Evans, B.A.J., Evidence for adenosine receptor regulation of osteogenesis versus adipogenesis in mesenchymal stem cells, *Calcified Tissue International*, 83:9–10, 2008.

[67] Evans, B.A., Elford, C., Pexa, A., Francis, K., Hughes, A.C, Deussen, A., Ham, J., Human osteoblast precursors produce extracellular adenosine, which modulates their secretion of IL-6 and osteoprotegerin, *Journal of Bone and Mineral Research*, 21(2):228–236, February 2006.

[68] Mohamadnejad, M., Sohail, M.A., Watanabe, A., Krause, D.S., Swenson, E.S. and Mehal, W.Z., Adenosine inhibits chemotaxis and induces hepatocyte-specific genes in bone marrow mesenchymal stem cells, *Hepatology*, 51(3):963–973, March 2010.

[69] Katebi, M., Soleimani, M., Cronstein, B.N., Adenosine A_{2A} receptors play an active role in mouse bone marrow-derived mesenchymal stem cell development, *Journal of Leukocytes Biology*, 85(3):438–444, March 2009.

[70] Montesinos, M.C., Shaw, J.P., Yee, H., Shamamian, P., Cronstein, B.N., Adenosine A(2A) receptor activation promotes wound neovascularization by stimulating angiogenesis and vasculogenesis, *American Journal of Pathology*, 164(6):1887–1892, 2004.

[71] Pospisil, M., Hofer, M., Znojil, V., Vacha, J., Netikova, J., Hola, J., Radioprotection of mouse hemopoiesis by dipyridamole and adenosine monophosphate in fractionated treatment, *Radiation Research*, 142(1):16–22, April 1995.

[72] Hofer, M., Pospisil, M., Netikova, J., Znojil, V., Vacha, J., Hola, J., Radioprotective efficacy of dipyridamole and AMP combination in fractionated radiation regimen, and its dependence on the time of administration of the drugs prior to irradiation, *Physiological Research*, 44(2):93–98, 1995.

[73] Wang, L., Jacobsen, S.E., Bengtsson, A., Erlinge, D., P2 receptor mRNA expression profiles in human lymphocytes, monocytes and CD34[+] stem and progenitor cells, *BMC Immunology*, 5:16, August 2004.

[74] Riddle, R.C., Taylor, A.F., Rogers, J.R., Donahue, H.J., ATP release mediates fluid flow-induced proliferation of human bone marrow stromal cells, *Journal of Bone and Mineral Research*, 22(4):589–600, April 2007.

[75] Ke, H.Z., Qi, H., Weidema, A.F., Zhang, Q., Panupinthu, N., Crawford, D.T., Grasser, W.A., Paralkar, V.M., Li, M., Audoly, L.P., Gabel, C.A., Jee, W.S., Dixon, S.J., Sims, S.M., Thompson, D.D., Deletion of the P2X$_7$ nucleotide receptor reveals its regulatory roles in bone formation and resorption, *Molecular Endocrinology*, 17(7):1356–1367, July 2003.

[76] Li, J., Liu, D., Ke, H.Z., Duncan, R.L., Turner, C.H., The P2X$_7$ nucleotide receptor mediates skeletal mechanotransduction, *Journal of Biological Chemistry*, 280(52):42952–42959, December 2005.

[77] Riddle, R.C., Taylor, A.F., Rogers, J.R., Donahue, H.J., ATP release mediates fluid flow-induced proliferation of human bone marrow stromal cells, *Journal of Bone and Mineral Research*, 22(4):589–600, April 2007.

[78] Coppi, E., Pugliese, A.M., Urbani, S., Melani, A., Cerbai, E., Mazzanti, B., Bosi, A., Saccardi, R., Pedata, F., ATP modulates cell proliferation and elicits two different electrophysiological responses in human mesenchymal stem cells, *Stem Cells*, 25(7):1840–1849, July 2007.

[79] Ichikawa, J., Gemba, H., Cell density-dependent changes in intracellular Ca^{2+} mobilization via the P2Y$_2$ receptor in rat bone marrow stromal cells, *Journal of Cellular Physiology*, 219(2):372–381, May 2009.

[80] Kawano, S., Otsu, K., Kuruma, A., Shoji, S., Yanagida, E., Muto, Y., Yoshikawa, F., Hirayama, Y., Mikoshiba, K., Furuichi, T., ATP autocrine/paracrine signaling induces calcium oscillations and NFAT activation in human mesenchymal stem cells, *Cell Calcium*, 39(4):313–324, April 2006.

[81] Pansky, A., Roitzheim, B., Tobiasch, E., Differentiation potential of adult human mesenchymal stem cells, *Clinical Laboratory*, 53(1/2):81–84, 2007.

[82] Bunnell, B.A., Flaat, M., Gagliardi, C., Patel, B., Ripoll, C., Adipose-derived stem cells: Isolation, expansion and differentiation, *Methods*, 45(2):115–120, June 2008.

[83] Tobiasch E., Adult Human Mesenchymal Stem Cells as Source for Future Tissue Engineering, *Forschungsspitzen und Spitzenforschung*, 2009, V, pp329-338.

[84] Zuk P.A., Zhu M., Ashjian P., De Ugarte D.A., Huang J.I., Mizuno H., Alfonso Z.C., Fraser J.K., Benhaim P., Hedrick M.H., Human adipose tissue is a source of multipotent stem cells, *Molecular Biology Cell*, 13(12):4279–4295, 2002.

[85] Zippel, N., Limbach, C.A., Ratajski, N., Urban, C., Pansky, A., Luparello, C., Kassack, M.U., Tobiasch, E., Purinergic receptors influence the differentiation of human mesenchymal stem cells, *Stem Cells & Development*, July 2011.

[86] Scholze, N.J., Zippel, N., Müller, C.A., Pansky, A., Tobiasch, E., P2X and P2Y receptors in human mesenchymal stem cell differentiation, *Tissue Engineering Part A*, 15(3):677, March 2009.

[87] Tobiasch, E., Zippel, N., Scholze, N.J., Limbach, C., Urban, C., Luparello, C., Pansky, A., Kassack, M.U., P2 receptors influence differentiation in human mesenchymal and ectomesenchymal stem cells, *Purinergic Signal.*, 13(8):534–538, August 2010.

[88] Park, K.S., Kim, Y.S., Kim, J.H., Choi, B.K., Kim, S.H., Oh, S.H., Ahn, Y.R., Lee, M.S., Lee, M.K., Park, J.B, Kwon, C.H., Joh, J.W., Kim, K.W. and Kim, S.J., Influence of human allogenic bone marrow and cord blood-derived mesenchymal stem cell secreting trophic factors on ATP (adenosine-50-triphosphate)/ADP (adenosine-50-diphosphate) ratio and insulin secretory function of isolated human islets from cadaveric donor, *Transplantation Proceedings*, 41(9):3813–3818, November 2009.

[89] Orciani, M., Mariggiò, M.A., Morabito, C., Di Benedetto, G., Di Primio, R., Functional characterization of calcium-signaling pathways of human skin-derived mesenchymal stem cells, *Skin Pharmacology and Physiology*, 23(3):124–132, 2010.

[90] Zippel, N., Schulze, M., Tobiasch, E., Biomaterials and mesenchymal stem cells for regenerative medicine, *Recent Patents on Biotechnology*, 4(1):1–22, January 2010.

[91] Vivas, D., Angiolillo, D.J., Platelet P2Y$_{12}$ receptor inhibition: An update on clinical drug development, *American Journal of Cardiovascular Drugs*, 10(4), 217–226, 2010.

8

Stem Cell Delivery into the Myocardial Wall

Giancarlo Forte, Stefano Pietronave, Francesca Pagliari,
Stefania Pagliari, Eugenio Magnani, Giorgia Nardone,
Enrico Traversa, Maria Prat, Marilena Minieri and Paolo Di Nardo

*Laboratorio Cardiologia Molecolare & Cellulare, Dipartimento Medicina Interna,
Università di Roma Tor Vergata, Roma, Italy; e-mail: dinardo@uniroma2.it*

Abstract

Stem cells have been proposed as a powerful tool for the treatment of cardiac diseases. Indeed, the identification of a small number of progenitor cells virtually within every tissue of the body opened new perspectives to setup specific therapeutic protocols for cardiac diseases. To this end, a number of pre-clinical and clinical trials have been already completed, in which different stem cell subsets were directly injected into the myocardium or delivered via bloodstream to the heart. Despite the evidence of little, but significant improvements in cardiac outcome, the results demonstrated that no or few cells were retained within the host tissue few weeks after cell administration. The mild beneficial effects on the host tissue were thus ascribable to paracrine factors released by the implanted cells before they are washed away or removed by the immune system.

Such results suggest that additional efforts are needed to setup efficient systems to deliver stem cells to the injured site. Among others, cardiac tissue engineering concepts could represent an efficient and cost/effective solution. However, their application requires cutting edge technologies to fabricate synthetic and hybrid scaffolds to be used as cell delivery systems. In this respect, particular attention must be paid to determine the role of the scaffold surface physical, mechanical and chemical properties and their effects on stem cells.

P. Di Nardo (Ed.), Adult Stem Cell Standardization, 99–110.

Keywords: cardiac stem cells, scaffold, delivery system, tissue engineering.

8.1 Introduction

In healthy human hearts, only 10–20% of the total cells are contractile cardiomyocytes and, at the age of 25 years, no more than 1% of them are annually substituted by progenitor cells, this percentage reducing to less than 0.5% at the age of 75. In total, less than 50% of cardiomyocytes are renewed within the normal human lifespan [4]. The incapability to provide an adequate maintenance of the cardiac muscle, very likely, makes it difficult to formulate decisive treatments for heart diseases. Indeed, cardiac diseases remain a predominant cause of mortality and morbidity in industrialized countries and, despite the recent advancements achieved in their treatment, end-stage heart failure management still relies on organ transplantation as the only approach, although 30–40% of patients suffer from immune rejection during the first year post-transplant [14]. To circumvent the limits of the therapeutic strategies, current post-infarction myocardial revascularization protocols include the administration of bone marrow mesenchymal stem cells, while a number of clinical trials have been performed or are in progress in which different stem cell subsets are implanted in the damaged tissue by means of surgical techniques. However, the results of such trials are still controversial. In fact, when autologous skeletal myoblasts had been injected into the heart of patients suffering from ischemic cardiomyopathy, the limited functional improvement was associated with severe arrhythmic events requiring the adoption of pacemakers/defibrillators [19]. On the other side, intracoronary administration of bone marrow mesenchymal stem cells resulted in modest improvements in the cardiac contractile function in patients with dilated cardiomyopathy [11]. Among multiple potential causes, the lack of efficient cell therapy strategies for cardiac diseases is determined by the rudimentary protocols so far applied to select appropriate cell subsets and to generate new vessels and contractile cardiomyocytes. Besides, the route by which stem cells should be delivered to the target organ is still debated. The solution of such problems requires additional efforts in basic research to study the processes leading to stem cell differentiation as well as technological advancements to setup efficient protocols to implant the cells.

In principle, adult stem cells could be isolated from patients' own tissues and expanded *in vitro* to reach a coherent number by means of well-known techniques including the adoption of GMP protocols to avoid cell contamination (Figure 8.1). Nonetheless, a number of issues should be challenged

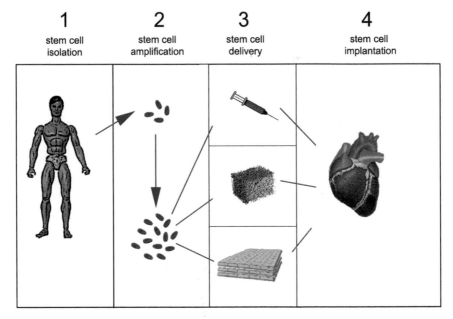

Figure 8.1 Adult stem cells can be harvested and purified from patient body, amplified in culture and delivered to the injured heart by direct injection, using a porous scaffold or by scaffoldless technology.

before safe procedures to manipulate stem cells *in vitro* for cardiac transplant can be setup. In fact, adult stem cells cannot be administered to patients with hereditary diseases and malignant transformation is likely to occur in *ex vivo* expanded cells when standard culture medium is adopted [13, 23]. Alternatively, stem cells could undergo senescence after a short number of passages *in vitro* [30] and induce a certain degree of host immune reaction. Finally, the use of animal-derived supplements could discourage the use of stem cells for cardiac cell therapy. Advantages in the administration of autologous stem cells could be the absence of immune rejection and the consequent need for critical immune-suppressive drugs.

8.2 Methods to Deliver Stem Cells to the Injured Heart

Different stem cell delivery methods have been proposed so far, but all of them suffer from a number of drawbacks.

8.2.1 Injection

Stem cell direct *intramural injection* has been widely applied generating contrasting results. For example, pre-clinical studies performed on experimental animals demonstrated that, although a certain extent of cardiac repair was achieved when bone marrow Stro-3+ perivascular cells were implanted *in vivo*, cells vanished from the application site within few days [8]. In other reports, when Sca-1+ cardiac resident stem cells were injected in infarction border zone, a modest but significant improvement in cardiac function was reported, with evidence of cell engraftment and differentiation [28]. Finally, bone marrow-derived c-kit+ cells were shown to repair entire ventricular areas massively engrafting and differentiating in contractile and vascular figures *in vivo* [27]. It is interest to note that independent groups already demonstrated that c-kit+ bone marrow cells fail to acquire contractile phenotype when implanted in diseased myocardium [2, 24]. Such discrepancies are not surprising since different stem cell subsets or preparation protocols were probably used in these studies.

Alternative approaches to deliver stem cells into the heart wall are: (i) *intravenous*, (ii) through *coronary arteries* or even (iii) retrogradely through the *coronary sinus*. Results obtained in pre-clinical animal models showed that the first option is a minimally invasive. Nonetheless, injected cells could be sequestered in the lungs, liver or spleen as a result of blood flow and only a limited number of cells could be found in the infarcted area after transverse the arterial or capillary wall. An intracoronary approach allows for selective and more concentrated delivery in heart ischemic and border zone regions, although cell are required to migrate through the endothelium into the myocardium. Cell delivery through a catheter placed into the coronary sinus allows for a homogeneous cell distribution, but still endothelial transmigration is required [6].

Finally, an interesting attempt with stem cells being injected into the *pericardial cavity* has been proposed. A high number of cells could be deposited and retained within the pericardial cavity, even if migration across the visceral pericardium is required (Table 8.1).

8.2.2 Injectable Scaffolds

Injectable scaffolds of biocompatible microspheres with dimensions adequate to surpass the capillary barrier are considered as a promising tool for stem cell delivery to damaged myocardium. Such an approach has already been used in the treatment of neurological diseases *in vivo* [21]. Recently, our laboratory

Table 8.1 Injection methods to deliver stem cells to diseased myocardium.

method	advantages	drawbacks
Intravenous delivery	No invasive technique	Cells can be sequestered in lung, liver, spleen
Intracoronary delivery	No risk of systemic delivery Direct delivery to the target site	Few cells delivered
Intramyocardial delivery	Direct delivery	Risk of perforation
Retrograde coronary sinus delivery	Homogeneous cell delivery	Endothelial wall transmigration required
Intrapericardial delivery	Large number of cells delivered	Visceral pericardium transmigration required

explored the possibility of using injectable scaffolds by interfacing murine mesenchymal (mMSC) and cardiac stem (mCSC) cells with poly-lactic acid (PLA) microspheres having a diameter of 30 and 100 μm. Preliminary *in vitro* experiments demonstrated that such cells can be grown onto PLA microspheres while preserving their phenotype, but the formation of cell clumps can hamper the application of this technique [12]. A different approach has been recently proposed to deliver human mesenchymal stem cells (hMSC) to diseased myocardium. *In vitro* experiments showed that hMSC could survive after encapsulation in RGD-modified alginate microsphere (diameter: 200–700 μm), proliferating and migrating through the porous material. When intramyocardially injected in a rat model of heart infarction obtained by the ligation of the left anterior descendant coronary (LAD), cell-loaded alginate microspheres promoted angiogenesis and prevented LV negative remodeling [29]. Nonetheless, few cells were found in the injection area after few days. Moreover, 10 weeks after the injection, microbead remains were still present within the host myocardium. The aspect of microbead resorption should thus be addressed before clinical perspectives could be foreseen.

8.2.3 Scaffold-Based Technology

The possibility of using biocompatible scaffolds to deliver stem cells to the injured heart has been explored by a number of independent research groups so far. The scaffolds proposed are fabricated with natural or synthetic materials, but, when designing cardiac-specific constructs, a number of requirements should be fulfilled. For example, it cannot be neglected that myocardial contractile function relies on the transmission of electrical and mechanical forces throughout a functional syncityum. So, the integrity of the tissue has to be preserved. For this reason, a cardiac-specific scaffold should comply with tissue architecture and, thus, be deformable enough to indulge and, if possible, sustain cardiac contraction. Moreover, as far as stem cell cardiac engraftment is concerned, scaffolds should be able to start at least cell alignment and commitment to favor stem cell electromechanical coupling with host tissue. In this respect, the use of Cerium Oxyde nanoparticles to modify the surface of poly-lactic acid films and to obtain a controlled nanorugosity appears intriguing [15]. In fact, far from being a noxious compound for stem cells, ceria was able to induce cardiac stem cell alignment and growth. Nonetheless, cardiac tissue is extremely complex and highly demanding in terms of blood supply and catabolite removal, so that porous scaffolds that could allow microvascular branches formation and oxygen perfusion are to be preferred. To fulfill such requirements, neonatal cardiomyocytes were seeded in Collagen I + Matrigel to produce Engineered Heart Tissue (EHT). Continuous contractile activity up to 1 week *in vitro* as well as cell survival and integration *in vivo* in syngeneic rat hearts were reported [32]. In another attempt, anisotropic accordion-like honeycomb scaffolds were prepared by excimer laser microablation using poly(glycerol sebacate) as an elastomeric tool to mimic anisotropic cardiac muscle stiffness distribution [9]. Although such scaffolds promote neonatal rat cardiomyocyte alignment and contraction, the response of stem or progenitor cells and the *in vivo* testing have not been performed so far. The same material has been used to produce elastomeric patches on which human embryonic stem cell-derived cardiomyocytes were grown, showing that it is indeed possible to observe spontaneous beating activity *in vitro* up to 3 months [7]. Such patches were shown to be suitable as delivery systems and, when sutured in the absence of cells onto healthy rat left ventricle, they did not affect cardiac contractile activity. More basic studies were also conducted to study the ability of stem cells to interface with different synthetic and natural materials. In this respect, few research groups focused on the possibility to drive a certain extent of stem cell commitment through tailoring scaffold physical and chemical properties, independently of biological cues. As such, consensus on the ability of stem cells to sense substrate rugosity and elasticity has been reached [10]. Thus, in order to rule out the occurrence of spontaneous events of differentiation in implanted cells, the possibility to induce stem cell pre-commitment *in vitro* on scaffolds towards

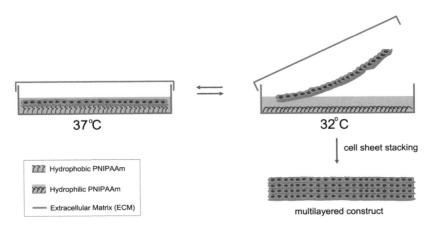

Figure 8.2 Fabrication of thick cardiac bio-substitutes by scaffoldless technology: Stem cells can be grown as a monolayer until they reach confluence onto poly-N-isopropylacrylamide (PNIPAAm)-coated thermo-responsive surfaces at 37°C. The temperature is lowered below 32°C, thus detaching a stem cell monolayer. The stacking of multiple monolayers can result in the generation of a multilayered thick construct to be transplanted (see also [17]).

a desired phenotype is being investigated. Indeed, Engler and collaborators compellingly demonstrated that it is possible to address stem cell fate by fine-tuning substrate elasticity as to match tissue-specific stiffness. Recently, this concept has been corroborated by our research group, showing that cardiac resident progenitors (Sca-1+ CPC) can be committed to cardiac phenotype by the mechano-structural signals arising from the matrix, but biological factors are needed to finalize the differentiation process [12, 26].

8.2.4 Scaffold-Free Technology

To overcome the problem of poor cell retention reported in cell injection experiments in the heart [8] and to avoid the release of possibly harmful scaffold byproducts, scaffold-free technology has been developed, in which cells are grown in a monolayer onto thermo-responsive surfaces and easily detached in the form of cell sheets by lowering the temperature [16]. Such technology takes advantage of the ability of polymers like the poly-N-isopropylacrylamide (PNIPAAm) to shift between the hydrophobic and hydrophilic status when the temperature ranges from 37 to 32°C. Cell sheets can therefore be serially stacked to obtain multilayered scaffoldless constructs (Figure 8.2).

Such an approach has already been applied to obtain cell sheets composed of rodent [22] and human differentiated [1] and undifferentiated cells [17]. Given the need for thick cardiac substitutes which should comply with cardiac muscle continuous contractility, thermo-responsive technology has been envisaged as a possible answer to the lack of heart donors. Pre-clinical trials performed onto experimentally infarcted animals demonstrated that, when a murine adipose-derived monolayer sheet is leant onto injured myocardium, it can be retained and help tissue repair [22]. More striking results were obtained when Sca-1+ cardiac progenitor cell-derived sheets were used [17]. Finally, an interesting approach has been recently proposed to deliver cardiac stem cells cultured in the form of cardiospheres to the injured heart: cardiospheres were embedded into a cardiac stromal cell-derived sheet obtained by using poly-lysine/collagen IV-coated dishes [31]. The formation of mature vessels as well as new cardiomyocytes *in vivo* was reported after 3 weeks.

8.2.4.1 Preparation and Transplantation of Human Cardiac Progenitor Cell Sheets

In this direction, our laboratory prepared multilayered cardiac patches starting from human cardiac resident Sca-1+ progenitor cells extracted from auricular biopsies. The cells obtained after informed consent werepurified according to c-kit expression. Surprisingly, c-kit expression was rapidly lost in culture, while Sca-1 antigen was found to be expressed at high levels throughout the culture (Figures 8.3a–c). The cells were uniformly fibroblastoid in shape and displayed vinculin adhesion processes (Figures 8.3d, e).

The cells displayed the ability to proceed to cardiac differentiation *in vitro* when co-cultured for 7 days with murine neonatal cardiomyocytes, as shown by GATA-4, connexin 43 and alpha sarcomeric actinin expression (Figure 8.4). Human cardiac progenitor cells were seeded onto PNIPAAm-coated dishes (Fischer Scientific) and grown until they reached confluence. The temperature was lowered below 32°C and a cell sheet detached. The cell sheet was implanted onto the left ventricle of an immunosuppressed mouse for 1 week.

At this point in time, the cell sheet (red cells in Figure 8.5) could still be found attached over the inner pericardial layer and some cells could be found inside the texture of murine myocardium.

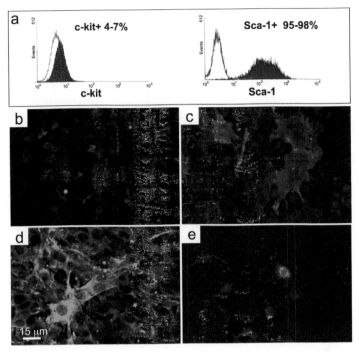

Figure 8.3 Fabrication of thick cardiac bio-substitutes by scaffoldless technology: Stem cells can be grown as a monolayer until they reach confluence onto poly-N-isopropylacrylamide (PNIPAAm)-coated thermo-responsive surfaces at 37°C. The temperature is lowered below 32°C, thus detaching a stem cell monolayer. The stacking of multiple monolayers can result in the generation of a multilayered thick construct to be transplanted (see also [17]).

8.3 Conclusions

The use of autologous stem cells as a tool to produce new cardiomyocytes is considered a promising approach for the set-up of innovative, minimally invasive therapies for cardiac diseases. Although significant advancements have been achieved in the purification and differentiation of stem cells, further efforts to translate basic knowledge to the bedside are required. In this context, the identification of safe, standardized protocols to deliver stem cells to the damaged site is a key step. For this reason, a consensus in scaffold fabrication protocols, cell manipulation and surgery maneuvers is needed to maximize cell retention in the organ while minimizing the impact of immune rejection.

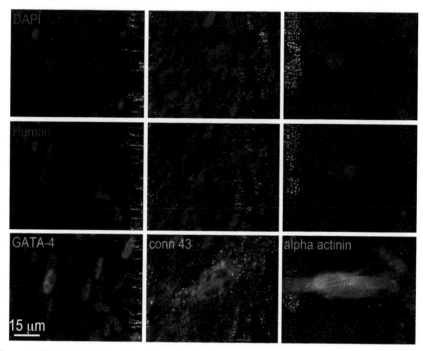

Figure 8.4 Human cardiac progenitor cell (hCPC) differentiation in co-culture with neonatal cardiomyocytes. hCPC were red-labelled with vital VYBRANT dye, while cardiac-specific antibodies (GATA-4, conn 43, alpha actinin) were visualized in green. Nuclei were counter-stained in blue (DAPI).

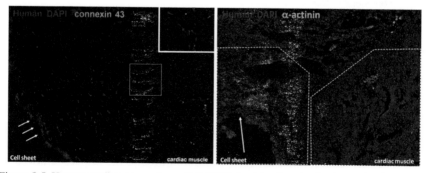

Figure 8.5 Human cardiac progenitor cell (hCPC) sheet implant *in vivo* in murine heart. One week after sheet implantation, human cells (red) migrated into murine.

References

[1] Arauchi A., Shimizu T., Yamato M., Obara T., Okano T. Tissue-engineered thyroid cell sheet rescued hypothyroidism in rat models after receiving total thyroidectomy comparing with non transplantation models. *Tissue Eng. Part A*, 15:3943–3949, 2009.

[2] Balsam L.B., Wagers A.J., Christensen J.L., Kofidis T., Weissman I.L., Robbins R.C. Haematopoietic stem cells adopt mature haematopoietic fates in ischaemic myocardium. *Nature*, 428:668–673, 2004.

[3] Bel A., et al. Composite cell sheets: A further step toward safe and effective myocardial regeneration by cardiac progenitors derived from embryonic stem cells. *Circulation*, 122:S118–123, 2010.

[4] Bergmann O., et al. Evidence for cardiomyocyte renewal in humans. *Science*, 324:98–102, 2009.

[5] Blin G., et al. A purified population of multipotent cardiovascular progenitors derived from primate pluripotent stem cells engrafts in postmyocardial infarcted nonhuman primates. *J. Clin. Invest.*, 120:1125–1139, 2010, doi: 10.1172/JCI40120.

[6] Bui Q.T., Gertz Z.M., Wilensky R.L.. Intracoronary delivery of bone-marrow-derived stem cells. *Stem Cell Res. Ther.*, 1:29–35, 2010.

[7] Chen Q.Z., et al. An elastomeric patch derived from poly(glycerol sebacate) for delivery of embryonic stem cells to the heart. *Biomaterials*, 31:3885–3893, 2010.

[8] Dixon J.A., et al. Mesenchymal cell transplantation and myocardial remodeling after myocardial infarction. *Circulation*, 120:S220–S229, 2009.

[9] Engelmayr G.C. Jr, Cheng M., Bettinger C.J., Borenstein J.T., Langer R., Freed L.E. accordion-like honeycombs for tissue engineering of cardiac anisotropy. *Nat. Mater.*, 7:1003–1010, 2008.

[10] Reilly G.C., Engler A.J. Intrinsic extracellular matrix properties regulate stem cell differentiation. *J. Biomech.*, 43:55–62, 2010.

[11] Fischer-Rasokat U., et al. A pilot trial to assess potential effects of selective intracoronary bone marrow-derived progenitor cell infusion in patients with nonischemic dilated cardiomyopathy: Final 1-year results of the transplantation of progenitor cells and functional regeneration enhancement pilot trial in patients with nonischemic dilated cardiomyopathy. *Circ. Heart Fail.*, 2:417–423, 2009.

[12] Forte G., et al. Criticatility of the biological and physical stimuli array inducing resident stem cell determination. *Stem Cells*, 26:2093–2103, 2008.

[13] Foudah D., Redaelli S., Donzelli E., Bentivegna A., Miloso M., Dalprà L., Tredici G. Monitoring the genomic stability of in vitro cultured rat bone-marrow-derived mesenchymal stem cells. *Chromosome Res.*, 17:1025–1039, 2009.

[14] Hertz M.I., et al. Scientific registry of the International Society for Heart and Lung Transplantation. *J. Heart Lung Transplant*, 28:989–1049, 2009.

[15] Mandoli C., Pagliari F., Pagliari S., Forte G., Di Nardo P., Licoccia S., Traversa E. Stem cell aligned growth induced by CeO2 nanoparticles in PLGA scaffolds with improved bioactivity for regenerative medicine. *Adv. Funct. Mater.*, 20:1617–1624, 2010.

[16] Masuda S., Shimizu T., Yamato M., Okano T. Cell sheet engineering for heart tissue repair. *Adv. Drug Deliv. Rev.*, 60:277–285, 2008.

[17] Haraguchi Y., Shimizu T., Yamato M., Kikuchi A., Okano T. Electrical coupling of cardiomyocyte sheets occurs rapidly via functional gap junction formation. *Biomaterials*, 27:4765–4774, 2006.

[18] Menasché, P. Stem cell therapy for heart failure: Are arrhythmias a real safety concern? *Circulation*, 119:2735–2740, 2009.

[19] Menasché P. Skeletal myoblasts for cardiac repair: Act II? *J. Am. Coll. Cardiol.*, 52:1881–1883, 2008.

[20] Menasché P., et al. The Myoblast Autologous Grafting in Ischemic Cardiomyopathy (MAGIC) trial: First randomized placebo-controlled study of myoblast transplantation. *Circulation*, 117:1189–1200, 2008.

[21] Menei P., Montero-Menei C., Venier M.C., Benoit J.P. Drug delivery into the brain using poly(lactide-co-glycolide) microspheres. *Expert Opin. Drug Deliv.*, 2:363–376, 2005.

[22] Miyahara Y., et al. Monolayered mesenchymal stem cells repair scarred myocardium after myocardial infarction. *Nat. Med.*, 12:459–465, 2006.

[23] Momin E.N., Vela G., Zaidi H.A., Quiñones-Hinojosa A. The oncogenic potential of mesenchymal stem cells in the treatment of cancer: Directions for future research. *Curr. Immunol. Rev.*, 6:137–148, 2010.

[24] Murry C.E., et al. Haematopoietic stem cells do not transdifferentiate into cardiac myocytes in myocardial infarcts. *Nature*, 428:664–668, 2004.

[25] Okura H., et al. Cardiomyoblast-like cells differentiated from human adipose tissue-derived mesenchymal stem cells improve left ventricular dysfunction and survival in a rat myocardial infarction model. *Tissue Eng. Part C Methods*, 16:417–425, 2010.

[26] Pagliari S., et al. Cooperation of biological and mechanical signals in cardiac progenitor cell differentiation. *Adv. Mater.*, 2010, doi: 10.1002/adma.201003479.

[27] Orlic D., Kajstura J., Chimenti S., Jakoniuk I., Anderson S.M., Li B., Pickel J., McKay R., Nadal-Ginard B., Bodine D.M., Leri A., Anversa P. Bone marrow cells regenerate infarcted myocardium. *Nature*, 410:221–229, 2001.

[28] Dixon J.A., Gorman R.C., Stroud R.E., Bouges S., Hirotsugu H., Gorman J.H. 3rd, Martens T.P., Itescu S., Schuster M.D., Plappert T., St John-Sutton M.G., Spinale F.G. Mesenchymal cell transplantation and myocardial remodeling after myocardial infarction. *Circulation*, 120:S220–S229, 2009.

[29] Yu J., et al. The use of human mesenchymal stem cells encapsulated in RGD modified alginate microspheres in the repair of myocardial infarction in the rat. *Biomaterials*, 31:7012–7020, 2010.

[30] Vacanti V., Kong E., Suzuki G., Sato K., Canty J.M., Lee T. Phenotypic changes of adult porcine mesenchymal stem cells induced by prolonged passaging in culture. *J. Cell Physiol.*, 194–201, 2005.

[31] Zakharova L., Mastroeni D., Mutlu N., Molina M., Goldman S., Diethrich E., Gaballa M.A. Transplantation of cardiac progenitor cell sheet onto infarcted heart promotes cardiogenesis and improves function. *Cardiovasc. Res.*, 87:40–49, 2010.

[32] Zimmermann W.H., et al. Engineered heart tissue grafts improve systolic and diastolic function in infarcted rat hearts. *Nat. Med.*, 12:452–458, 2006.

9

Cellular Therapies and Regenerative Medicine Strategies for Diabetes

Camillo Ricordi[1] and Juan Dominguez-Bendala[2]

Diabetes Research Institute, University of Miami Miller School of Medicine, Miami, Florida, USA; e-mail: {cricordi, jdominguez2}@med.miami.edu

Abstract

The prospects of developing stem cell-based therapies for the treatment of insulinopenic diabetes have been stoked by recent progress in islet transplantation. A decade from the breakthrough implementation of steroid-free immunosuppression methods that first allowed for long-term survival of the grafts (the Edmonton protocol), nowadays the rate of success of this approach stands comparison to that reported for whole pancreas transplantation. If effectively coaxed into insulin-positive derivatives, stem cells of different origins could offer a self-renewable alternative to islets, for which there is not a sustainable supply. Our chapter reviews the latest advances in the development of clinically relevant methods for the generation of new beta cells from both embryonic and adult stem cells, as well as the experimental conversion of non-endocrine-pancreatic tissues (chiefly liver and pancreatic acini) into insulin-producing beta cells.

Keywords: stem cells, reprogramming, diabetes, islet transplantation.

9.1 Introduction

Difficult as it may be to believe today, a simple search in PubMed reveals that the concept of regenerative medicine not never used in the scientific literature

P. Di Nardo (Ed.), Adult Stem Cell Standardization, 111–126.

prior to 2000. Just a decade after that, the notion that we could use stem cells to treat conditions for which there is no known cure has pervaded the entire medical field. Even if the frantic pace of research has not been fully translated yet into clinical therapies, their advent in the near future is widely thought to be nothing short of inevitable.

For many reasons, type 1 diabetes appears to be among the best positioned diseases to be targeted by stem cell-based therapies. This is so despite the well-known resistance of most stem cell types studied thus far to become insulin-producing beta cells. Indeed, this is a cellular fate difficult to get to *in vitro*, especially when compared to other popular destinations such as neural or cardiomyocyte phenotypes. However, the fact that there is already a cell therapy for the disease represents a very significant advantage. If the future holds a cure for Alzheimer's disease, for instance, we can be certain that researchers will have to go through all the stages of a painstaking learning curve in order to troubleshoot the many obstacles that will be found in the way. Islet transplantation, in contrast, has already paved the way for the pro-spective stem cell treatments of the XXI century for diabetes. After more than two decades of honing this experimental approach (in which the islets that contain the beta cells of the pancreas are isolated and transplanted in the liver of the patient), islet transplantation offers now diabetes reversal rates and length of graft function that are on par with the current standard of whole pancreas transplantation – minus the risks associated with major surgery. This biomedical feat has allowed hundreds of patients to remain either off insulin or with a much better metabolic control years after the islets were implanted. Because the principle has been established, it is expected that when a self-renewable source (be it of adult or embryonic origin) is shown to effectively and safely adopt the beta cell phenotype in defined conditions, we will be in a position to swiftly adapt current protocols switching islets by stem cells – thus readily extending the benefits of islet transplantation to millions of patients [1]. This chapter will review the numerous strategies along these lines that have been conceived over the past few years, many of which are already finding their way to the clinical practice.

9.2 The Problem

Type 1 diabetes is the result of an autoimmune response against the pan-creatic beta cells, which maintain glucose homeostasis by secreting insulin into the bloodstream. Although the predisposition to develop the disease has a well studied genetic component, much remains to be learnt about the trigger, which is thought to be environmental. The autoimmune response will progressively attack the beta cells of the islets. After a critical mass has been destroyed, clinical symptoms will occur. These are characterized by the inability to regulate blood sugar levels. Chronic insulin administration is a life-saving procedure, but cannot maintain glucose concentrations within a physiological range. Sustained deviations over the years result in clinical complications that chiefly involve a wealth of circulatory problems. Blindness, kidney failure, ulcers and amputations are typical outcomes. The advent of the insulin pump (which delivers insulin in a continuous fashion through a subcutaneous catheter) has afforded patients a greater flexibility for the management of the disease, but it has in no way represented a solution. Whole pancreas transplantation is an effective treatment, but poses all the risks inherent to major surgery [2].

9.3 First-Generation Cell Therapies

Islet transplantation was conceived as an alternative treatment that combined the benefits of whole pancreas transplantation and the low risk of an outpatient procedure [3–8]. The process basically entails the separation of the endocrine component of the pancreas (the islets) from the rest of the acinar tissue, which is not required for glucose homeostasis. By doing so, and considering that islets account for only 1–2% of the total mass of the organ, clinicians end up with a minimal amount of tissue that can be ectopically implanted. Following the enzymatic/mechanical dissociation of the islets [9], centrifugation enriches for the fractions (layers) that contain the higher proportion of islets. In its most common embodiment, this therapy is administered by infusing the islets intraportally by gravity in the liver of the patient. As they course through progressively smaller blood vessels, the islets eventually embolize and remain permanently in place [10]. Revascularization, which ensues within a few weeks [11–13], allows for the long-term engraftment of these islets.

To this date, IBMIR (instant blood-mediated inflammatory reaction) remains a major factor in the rapid demise of a large percentage of freshly transplanted islets [14]. For more than a decade after the first clinical trials, this islet cell loss was compounded by the effects of the very same steroid-based immunosuppression agents that were used to prevent rejection. The introduction of glucocorticoid-free protocols in 2000 was a breakthrough that allowed for the long-term survival and function of transplanted islets [15], albeit not indefinitely [16]. However, the cumulative effect of progressive refinements in immunosuppression has finally resulted in engraftment and function rates that stand comparison with those seen in whole pancreatic transplantation [17].

For all the success of islet transplantation over the last decade, it is not uncommon to see that insulin independence still requires more than one donor [15, 18, 19]. The limited availability of pancreata suitable for processing and transplantation has prevented the widespread use of this therapy. There is a consensus that, in its current form, the rate of progress in islet transplantation has already plateaued.

9.4 Next-Generation Cell Therapies

9.4.1 Pluripotent Stem Cells

Pluripotent stem cells are generally defined as those with the capability of turning into derivatives of all three germ layers (endoderm, ectoderm and mesoderm), both *in vivo* and *in vitro*. Because of our 30-plus year experience with their kind (first from mice [20, 21], and since 1998 from humans [22]), embryonic stem (ES) cells are the best known and most widely studied of them all. However, there are several other pluripotent stem cell types, including embryonic germ (EG) cells and the most recently described induced pluripotent stem (iPS) cells. Unlike ES and EG cells, which can be considered a self-perpetuating snapshot of specific phases of embryonic development, the latter have no correspondence in nature because they are generated *ex vivo* by the artificial reprogramming of non-pluripotent somatic cells [23–25]. However, their behaviour and phenotype are, to all practical purposes, indistinguishable from those of "natural" pluripotent stem cells. They also have the added advantage that they could be derived from the prospective patient, thus overcoming allorejection (but not autoimmunity) [1]. The possibility of turning self-renewable, pluripotent cells into insulin-producing cells was recognized from very early on after the isolation of human ES cells. However, progress at recapitulating beta cell differentiation *in vitro* has been slow and plagued by numerous deviations that ended up being dead ends [26–28]. Even when the conditions for the differentiation of definitive endoderm were finally defined [29], other problems such as tumorigenesis [30–32] and lack of efficiency [32, 33] were identified.

It is now generally believed that the protocol originally reported by Novocell (now ViaCyte) [32, 34, 35] represents the state-of-the-art. Indeed, preclinical work on this method has been deemed promising enough to warrant the award of a $20 million-plus grant by the California Institute of Regenerative Medicine (CIRM) with the goal of speeding up the development of clinical therapies.

Chief amongst the concerns regarding the translation of these preclinical reports is safety. This is due to the fact that limitations in differentiation efficiency are overcome by transplanting the cells at an immature stage [34]. The *in vivo* microenvironment provides the necessary cues for these cells to mature into functional beta cells, but at the expense of an increased risk of tumorigenesis. The thorough screening for undifferentiated cellular byproducts [36], the stable integration of suicide genes that would be activated only in the cells that kept dividing after differentiation [37–41], and the use of encapsulation/physical containment techniques that would both immunoisolate the graft and prevent the spread of a tumor [42–45] are some of the approaches that have been proposed to reduce this risk to clinically acceptable levels. The latter is the strategy favoured by ViaCyte, whose scientists contend that this system would allow not only for the physical immunoisolation of allogeneic grafts, but also for the prompt destruction of any potential undifferentiated escapees by the recipient's fully competent immune system.

9.4.2 Adult Stem Cells

Are there stem cells in the pancreas? If there were, many argue, there would be no point in exploring the much more controversial – and potentially dangerous – pluripotent stem cells. If these cells existed, perhaps it would even be possible to 'reactivate' them in diabetic patients without the need for more invasive interventions. Otherwise, they could be biopsied, expanded *ex vivo* and re-implanted in the patient. Being autologous, rejection (other than autoimmunity) would not be an issue.

But is this a real possibility? Decades of research have not offered a conclusive answer to the question of whether the pancreas is home to organ-specific stem cells with the above potential. Investigators from perhaps the most influential camp claim that pancreatic stem cells do exist, and they reside in the ductal system [46]. In support of this contention is the common observation that either single beta cells or small clusters of them are often seen in the proximity of ducts. While these findings are undisputable, skeptics point out that the lack of 3D time-lapse microphotographs documenting this phenomenon makes it impossible to determine whether these cells are mini-islets in the process of formation from the duct or just the result of a random selection of section planes (i.e., the "tip" of a bigger islet). Still, a considerable amount of *in vivo* and *in vitro* studies show that ductal cells can be induced under specific circumstances to express endocrine cell markers, including insulin [47–54]. Also, ontological equivalences with the liver (where endodermal stem cells have been found in peribiliary glands that are often associated to ductal-like structures [55, 56]) are suggestive that the ducts may contain pancreatic progenitor cells after all.

The opposite camp's arguments are solidly rooted in a series of breakthrough reports in the mid-2000s [57–59] that demonstrated that the normal turnover and regeneration of beta cells in the mouse pancreas does not involve the activation of a putative stem cells, but rather the division of pre-existing adult beta cells. This was true of several models of regeneration, including partial pancreatectomy and pregnancy.

As it is often the case in this ever-evolving field, the debate was not settled by these elegant reports. Indeed, additional lineage-tracing experiments seemed to confirm that the ducts could also give rise to new beta cells [60]; and that partial duct ligation (another model of beta cell regeneration not previously explored in this context) resulted in a reactivation of pancreatic stem cells (Pdx1+/Ngn3+) that were widely thought to be extinguished shortly before the completion of embryonic development. Paradoxically, these 'resuscitated' fetal stem cells were associated neither with ducts nor with islets.

In summary, the jury is still out. Since nobody has been able yet to isolate and expand endocrine stem cells from the fully developed pancreas, let alone find the right stimuli to activate them *in vivo*, there has been a very active search for adult stem cells in other places. The bone marrow has been the long-standing favourite of those who argue that there is a universal, centralized 'repair kit', one amenable to organ-targeted reactivation upon local injury during adulthood. Indeed, the migration of exogenous bone marrow cells to selected tissues has been often reported in numerous species, including humans [61–75]. Although the self-repair notion was shaken in the early 2000s by several reports suggesting that presumptive regeneration could actually be explained by the fusion between resident and immigrant cells [72, 76, 77], a series of creative experiments involving the Cre-mediated labelling of putative fusion events seemed to rule out this possibility in a model of bone marrow colonization of islets [78]. However, two studies published shortly thereafter found little or no evidence of bone marrow conversion into pancreatic beta cells [79, 80]; and a third one [81] shed additional light on the subject by proving that bone marrow-derived endothelial progenitor cells could indirectly contribute to islet regeneration by improving vascularization and/or optimizing the local microenvironment conditions.

Hematopoietic stem cells (HSCs) from either the bone marrow or the umbilical cord blood could also be used in transplantation settings. These *bona fide* stem cells remain active during adulthood and can be banked so that HLA representation is maximized for prospective recipients. Thus far, their main application in the context of diabetes has not been for regeneration but rather to reset the immunological clock of diabetes [82–85], a very promising intervention aimed at tackling the root of the problem. However, some groups have already started to administer these cells into the pancreas of the patient through interventional radiology techniques, in the hope that they will induce pancreatic islet regeneration. This approach, combined with hyperbaric oxygen treatment, has already shown therapeutic potential for type 2 diabetes in preliminary trials [86].

Still, the workhorse of all adult stem cells remains the mesenchymal stem cell (MSC). These cells are defined by their ability to adhere to and rapidly divide in plastic, and can be isolated from many tissues, including not only the above-mentioned hematopoietic sources (umbilical cord and bone marrow) but also the pancreas and the adipose tissue [87–103]. Being the most easily accessible source, the latter has already shown tremendous potential for banking [104]. Indeed, adipose-derived MSCs obtained from eyelid plastic surgery procedures have garnered some attention as a potential substrate for beta-like cell differentiation [105]. However, in general the differentiation of MSCs into endodermal, beta-like cells has been very challenging. The reason for this is that, unlike pluripotent stem cells, MSCs are thought to be already committed along the mesodermal lineage. With very few exceptions (all of which share the common theme that the cells used may actually be more ES-like than mesenchymal [106]) success at reversing diabetes in animal models has been only modest at best [107–110]. Another important hurdle has been the lack of a "consensus" differentiation method, which makes it all the more difficult to compare the relative potency of MSCs from different origins. These difficulties are compounded by the observation that even different clones from the same tissue and donor exhibit different properties [111].

9.4.3 Reprogramming

Novel reprogramming/transdifferentiation strategies aim at exploiting the observation that some adult tissues might be converted into others under the appropriate circumstances. Undoubtedly because of their close ontogenetic ties [112–120], the pancreas and the liver have been shown to interconvert with relative ease [121–126]. One of the first studies to explore the molecular determinants behind this phenomenon made use of the pancreatic master gene Pdx1 [127, 128]. Ectopic Pdx1 expression in the liver of diabetic mice was shown to induce the activation of several beta cell-specific genes [129], leading to a sustained reduction in blood sugar levels that persisted long after the virus used to deliver the gene had been cleared from the system [130, 131]. Many other groups expanded these original findings by looking at polygenic approaches [132–139], which offered a glimpse of the possibilities but never materialized into full, *bona fide* beta cell transdifferentiation. Only a high-impact 2008 report by Zhou et al. [140] was able to shake out the relative stagnation of a field that seemed past its heyday by the second half of the decade. This group screened multiple permutations of key pancreatic transcription factors until they found an optimal combination (Pdx1, Ngn3 and MafA) that was proven to convert acinar tissue into beta-like cells *in vivo* as early as 3 days after the adenoviral-mediated injection of these three genes in the pancreatic parenchyma. As previously shown by others in the liver setting [130], only an initial trigger was necessary to induce irreversible reprogramming, even when the expression of the ectopic genes could no longer be detected. This observation is highly significant from the standpoint of developing clinical therapies, although *in vivo* reprogramming does not seem to be feasible in this context. Still, these experiments suddenly made us reconsider the seemingly uselessness of the copious amounts of acinar tissue that are thrown out after every islet isolation procedure. Indeed, this tissue is so plentiful (the ratio of exocrine to endocrine tissue is approximately 50:1 in the human pancreas) that *ex vivo* expansion might not even be necessary. Even a relatively low efficiency of conversion is likely to yield enough beta cell mass to treat multiple patients per organ. Novel means to induce reprogramming that do not involve the use of viruses (such as protein transduction [141, 142] or the most recent addition to the reprogramming armamentarium, modified mRNAs [143]) are the subject of very active research in this area.

9.5 Conclusion

Based on the steadfast progress that we have witnessed during the last decade, it is not inconceivable that we will see the development and widespread clinical use of stem cell-based treatments for diabetes before the end of the current one. Islet transplantation has been a trail blazer for new-generation therapies, whose implementation will not be hampered by the same painstaking trial-and-error tribulations that have shaped the progress of the former over the past two decades. In other words, to a large extent we know what to expect – and this knowledge can only be beneficial for the speedy translation of stem cell therapies.

Despite their tumorigenic potential, human ES will likely see the light of clinical trials within the next 5–8 years. The preclinical success of the ViaCyte protocol bodes well for the development of off-the-shelf treatments, provided that efficiency and safety concerns are properly addressed. Adult stem cells are still somewhat behind, notwithstanding the public perception otherwise. Although many clinics around the globe have already rushed to implement MSC or bone marrow-based therapies for type 1 diabetes (in many instances with a total disregard for scientific rigor and in a deregulated legal environment), evidence of success is anecdotal at best and a "gold standard" protocol of differentiation still needs to be defined. Only then will we be able to properly compare the potency of cells from all the different sources and unambiguously establish which one represents the best compromise between effectiveness and ease of procurement. In the meantime, we may also witness paradigm-shifting reprogramming techniques taking over the field and rapidly advancing all the way to clinical trials.

Acknowledgements

The authors state no conflict of interest and acknowledge the funding of the National Institutes of Health, The Juvenile Diabetes Research Foundation, the American Diabetes Association, the Foundation for Diabetes Research and the Diabetes Research Institute Foundation.

References

[1] Dominguez-Bendala, J. et al. Stem cell-derived islet cells for transplantation. *Curr. Opin. Organ. Transplant.*

[2] Burke, G.W. et al. Advances in pancreas transplantation. *Transplantation*, 77:62–67, 2004.

[3] Inverardi, L. et al. Islet transplantation: immunological perspectives. *Curr. Opin. Immunol.*, 15:507–511, 2003.

[4] Merani, S. et al. Current status of pancreatic islet transplantation. *Clin. Sci. (Lond)*, 110:611–625, 2006.

[5] Murdoch, T.B. et al. Methods of human islet culture for transplantation. *Cell Transplant*, 13:605–617, 2004.

[6] Pileggi, A. et al. Clinical islet transplantation. *Minerva Endocrinol.*, 31:219–232, 2006.

[7] Ricordi, C. et al. Clinical islet transplantation: Advances and immunological challenges. *Nat. Rev. Immunol.*, 4:259–268, 2004.

[8] Seung, E. et al. Induction of tolerance for islet transplantation for type 1 diabetes. *Curr. Diab. Rep.*, 3:329–335, 2003.

[9] Ricordi, C. et al. Automated method for isolation of human pancreatic islets. *Diabetes*, 37:413–420, 1988.

[10] Korsgren, O. et al. Optimising islet engraftment is critical for successful clinical islet transplantation. *Diabetologia*, 51:227–232, 2008.

[11] Carlsson, P.O. et al. Oxygen tension and blood flow in relation to revascularization in transplanted adult and fetal rat pancreatic islets. *Cell Transplant*, 11:813–820, 2002.

[12] Carlsson, P.O. et al. Low revascularization of experimentally transplanted human pancreatic islets. *J. Clin. Endocrinol. Metab.*, 87:5418–5423, 2002.

[13] Lammert, E. et al. Role of VEGF-A in vascularization of pancreatic islets. *Curr. Biol.*, 13:1070–1074, 2003.

[14] Johansson, H. et al. Tissue factor produced by the endocrine cells of the islets of Langerhans is associated with a negative outcome of clinical islet transplantation. *Diabetes*, 54:1755–1762, 2005.

[15] Shapiro, A.M. et al. Islet transplantation in seven patients with type 1 diabetes mellitus using a glucocorticoid-free immunosuppressive regimen. *N. Engl. J. Med.*, 343:230–238, 2000.

[16] Ryan, E.A. et al. Five-year follow-up after clinical islet transplantation. *Diabetes*, 54:2060–2069, 2005.

[17] Bellin, M.D. et al. Prolonged insulin independence after islet allotransplants in recipients with type 1 diabetes. *Am. J. Transplant*, 8:2463–2470, 2008.

[18] Froud, T. et al. Islet transplantation in type 1 diabetes mellitus using cultured islets and steroid-free immunosuppression: Miami experience. *Am. J. Transplant*, 5:2037–2046, 2005.

[19] Markmann, J.F. et al. Insulin independence following isolated islet transplantation and single islet infusions. *Ann. Surg.*, 237:741–749; discussion 749–750, 2003.

[20] Evans, M.J. et al. Establishment in culture of pluripotential cells from mouse embryos. *Nature*, 292:154–156, 1981.

[21] Martin, G.R. Isolation of a pluripotent cell line from early mouse embryos cultured in medium conditioned by teratocarcinoma stem cells. *Proc. Natl. Acad. Sci. USA*, 78:7634–7638, 1981.

[22] Thomson, J.A. et al. Embryonic stem cell lines derived from human blastocysts. *Science*, 282:1145–1147, 1998.

[23] Takahashi, K. et al. Induction of pluripotent stem cells from adult human fibroblasts by defined factors. *Cell*, 131:861–872, 2007.

[24] Takahashi, K. et al. Induction of pluripotent stem cells from mouse embryonic and adult fibroblast cultures by defined factors. *Cell*, 126:663–676, 2006.

[25] Yu, J. et al. Induced pluripotent stem cell lines derived from human somatic cells. *Science*, 318:1917–1920, 2007.

[26] Lumelsky, N. et al. Differentiation of embryonic stem cells to insulin-secreting structures similar to pancreatic islets. *Science*, 292:1389–1394, 2001.

[27] Baharvand, H. et al. Generation of insulin-secreting cells from human embryonic stem cells. *Dev. Growth Differ.*, 48:323–332, 2006.

[28] Hansson, M. et al. Artifactual insulin release from differentiated embryonic stem cells. *Diabetes*, 53:2603–2609, 2004.

[29] D'Amour, K.A. et al. Efficient differentiation of human embryonic stem cells to definitive endoderm. *Nat. Biotechnol.*, 23:1534–1541, 2005.

[30] Fujikawa, T. et al. Teratoma formation leads to failure of treatment for type I diabetes using embryonic stem cell-derived insulin-producing cells. *Am. J. Pathol.*, 166:1781–1791, 2005.

[31] Blyszczuk, P. et al. Expression of Pax4 in embryonic stem cells promotes differentiation of nestin-positive progenitor and insulin-producing cells. *Proc. Natl. Acad. Sci. USA*, 100:998–1003, 2003.

[32] D'Amour, K.A. et al. Production of pancreatic hormone-expressing endocrine cells from human embryonic stem cells. *Nat. Biotechnol.*, 2006.

[33] Jiang, J. et al. Generation of insulin-producing islet-like clusters from human embryonic stem cells. *Stem Cells*, 25:1940–1953, 2007.

[34] Kroon, E. et al. Pancreatic endoderm derived from human embryonic stem cells generates glucose-responsive insulin-secreting cells *in vivo*. *Nat. Biotechnol.*, 26:443–452, 2008.

[35] McLean, A.B. et al. Activin a efficiently specifies definitive endoderm from human embryonic stem cells only when phosphatidylinositol 3-kinase signaling is suppressed. *Stem Cells*, 25 29–38, 2007.

[36] Vogel, G. Cell biology. Ready or not? Human ES cells head toward the clinic. *Science*, 308:1534–1538, 2005.

[37] Tamada, K. et al. Molecular targeting of pancreatic disorders. *World J. Surg.*, 29:325–333, 2005.

[38] Wang, X.P. et al. Specific gene expression and therapy for pancreatic cancer using the cytosine deaminase gene directed by the rat insulin promoter. *J. Gastrointest. Surg.*, 8:98–108; discussion 106–108, 2004.

[39] Yazawa, K. et al. Current progress in suicide gene therapy for cancer. *World J. Surg.*, 26:783–789, 2002.

[40] Schuldiner, M. et al. Selective ablation of human embryonic stem cells expressing a "suicide" gene. *Stem Cells*, 21:257–265, 2003.

[41] Fareed, M.U. et al. Suicide gene transduction sensitizes murine embryonic and human mesenchymal stem cells to ablation on demand – A fail-safe protection against cellular misbehavior. *Gene Ther.*, 9:955–962, 2002.

[42] Beck, J. et al. Islet encapsulation: Strategies to enhance islet cell functions. *Tissue Eng.*, 13:589–599, 2007.

[43] Dang, S.M. et al. Controlled, scalable embryonic stem cell differentiation culture. *Stem Cells*, 22:275–282, 2004.

[44] Fort, A. et al. Biohybrid devices and encapsulation technologies for engineering a bioartificial pancreas. *Cell Transplant*, 17:997–1003, 2008.

[45] Orive, G. et al. History, challenges and perspectives of cell microencapsulation. *Trends Biotechnol.*, 22:87–92, 2004.

[46] Bonner-Weir, S. et al. A second pathway for regeneration of adult exocrine and endocrine pancreas. A possible recapitulation of embryonic development. *Diabetes*, 42:1715–1720, 1993.

[47] Bonner-Weir, S. et al. *In vitro* cultivation of human islets from expanded ductal tissue. *Proc. Natl. Acad. Sci. USA*, 97:7999–8004, 2000.

[48] Noguchi, H. et al. Induction of pancreatic stem/progenitor cells into insulin-producing cells by adenoviral-mediated gene transfer technology. *Cell Transplant*, 15:929–938, 2006.

[49] Yatoh, S. et al. Differentiation of affinity-purified human pancreatic duct cells to beta-cells. *Diabetes*, 56:1802–1809, 2007.

[50] Sharma, A. et al. The homeodomain protein IDX-1 increases after an early burst of proliferation during pancreatic regeneration. *Diabetes*, 48:507–513, 1999.

[51] Rosenberg, L. et al. Trophic stimulation of the ductular-islet cell axis: A new approach to the treatment of diabetes. *Adv. Exp. Med. Biol.*, 321:95–104; discussion 105–109, 1992.

[52] Gepts, W. Pathologic anatomy of the pancreas in juvenile diabetes mellitus. *Diabetes*, 14:619–633, 1965.

[53] Weaver, C.V. et al. Immunocytochemical localization of insulin-immunoreactive cells in the pancreatic ducts of rats treated with trypsin inhibitor. *Diabetologia*, 28:781–785, 1985.

[54] Martin-Pagola, A. et al. Insulin protein and proliferation in ductal cells in the transplanted pancreas of patients with type 1 diabetes and recurrence of autoimmunity. *Diabetologia*, 51:1803–1813, 2008.

[55] Cardinale, V. et al. Multipotent stem cells in the biliary tree. *Ital. J. Anat. Embryol.*, 115:85–90.

[56] Wang, Y. et al. Lineage restriction of human hepatic stem cells to mature fates is made efficient by tissue-specific biomatrix scaffolds. *Hepatology*.

[57] Dor, Y. et al. Adult pancreatic beta-cells are formed by self-duplication rather than stem-cell differentiation. *Nature*, 429:41–46, 2004.

[58] Teta, M. et al. Growth and regeneration of adult beta cells does not involve specialized progenitors. *Dev. Cell*, 12:817–826, 2007.

[59] Nir, T. et al. Recovery from diabetes in mice by beta cell regeneration. *J. Clin. Invest.*, 117:2553–2561, 2007.

[60] Xu, X. et al. Beta cells can be generated from endogenous progenitors in injured adult mouse pancreas. *Cell*, 132:197–207, 2008.

[61] Krause, D.S. et al. Multi-organ, multi-lineage engraftment by a single bone marrow-derived stem cell. *Cell*, 105:369–377. 2001.

[62] Ferrari, G. et al. Muscle regeneration by bone marrow-derived myogenic progenitors. *Science*, 279:1528–1530, 1998.

[63] Ferrari, G. et al. Myogenic stem cells from the bone marrow: A therapeutic alternative for muscular dystrophy? *Neuromuscul. Disord.*, 12, Suppl 1:S7–10, 2002.

[64] Gussoni, E. et al. Dystrophin expression in the mdx mouse restored by stem cell transplantation. *Nature*, 401:390–394, 1999.

[65] Orlic, D. et al. Transplanted adult bone marrow cells repair myocardial infarcts in mice. *Ann. NY Acad. Sci.*, 938:221–229; discussion 229–230, 2001.

[66] Orlic, D. et al. Bone marrow cells regenerate infarcted myocardium. *Nature*, 410:701–705, 2001.

[67] Orlic, D. et al. Mobilized bone marrow cells repair the infarcted heart, improving function and survival. *Proc. Natl. Acad. Sci. USA*, 98:10344–10349, 2001.

[68] Jackson, K.A. et al. Regeneration of ischemic cardiac muscle and vascular endothelium by adult stem cells. *J. Clin. Invest.*, 107:1395–1402, 2001.

[69] Lin, Y. et al. Origins of circulating endothelial cells and endothelial outgrowth from blood. *J. Clin. Invest.*, 105:71–77, 2000.

[70] Asahara, T. et al. Bone marrow origin of endothelial progenitor cells responsible for postnatal vasculogenesis in physiological and pathological neovascularization. *Circ. Res.*, 85:221–228, 1999.

[71] Theise, N.D. et al. Derivation of hepatocytes from bone marrow cells in mice after radiation-induced myeloablation. *Hepatology*, 31:235–240, 2000.

[72] Lagasse, E. et al. Purified hematopoietic stem cells can differentiate into hepatocytes *in vivo. Nat. Med.*, 6:1229–1234, 2000.

[73] Brazelton, T.R. et al. From marrow to brain: Expression of neuronal phenotypes in adult mice. *Science*, 290:1775–1779, 2000.

[74] Theise, N.D. et al. Liver from bone marrow in humans. *Hepatology*, 32:11–16, 2000.

[75] Korbling, M. et al. Hepatocytes and epithelial cells of donor origin in recipients of peripheral-blood stem cells. *N. Engl. J. Med.*, 346:738–746, 2002.

[76] Grompe, M. Bone marrow-derived hepatocytes. *Novartis Found Symp.*, 265:20–27; discussion 28–34, 92–97, 2005.

[77] Willenbring, H. et al. Myelomonocytic cells are sufficient for therapeutic cell fusion in liver. *Nat Med.*, 10:744–748, 2004.

[78] Ianus, A. et al. *In vivo* derivation of glucose-competent pancreatic endocrine cells from bone marrow without evidence of cell fusion. *J. Clin. Invest.*, 111:843–850, 2003.

[79] Choi, J.B. et al. Little evidence of transdifferentiation of bone marrow-derived cells into pancreatic beta cells. *Diabetologia*, 46:1366–1374, 2003.

[80] Lechner, A. et al. No evidence for significant transdifferentiation of bone marrow into pancreatic beta-cells *in vivo. Diabetes*, 53:616–623, 2004.

[81] Mathews, V. et al. Recruitment of bone marrow-derived endothelial cells to sites of pancreatic beta-cell injury. *Diabetes*, 53:91–98, 2004.

[82] Burt, R.K. et al. Randomized controlled trials of autologous hematopoietic stem cell transplantation for autoimmune diseases: the evolution from myeloablative to lymphoablative transplant regimens. *Arthritis Rheum.*, 54:3750–3760, 2006.

[83] Voltarelli, J.C. et al. Autologous nonmyeloablative hematopoietic stem cell transplantation in newly diagnosed type 1 diabetes mellitus. *Jama*, 297:1568–1576, 2007.

[84] Couri, C.E. et al. Autologous stem cell transplantation for early type 1 diabetes mellitus. *Autoimmunity*, 1, 2008.

[85] Couri, C.E. et al. Potential role of stem cell therapy in type 1 diabetes mellitus. *Arq. Bras. Endocrinol. Metabol.*, 52:407–415, 2008.

[86] Estrada, E., Valacchi, F., Nicora, E., Brieva, S., Esteve, C., Echevarria, L., Froud, T., Bernetti, K., Messinger Cayetano, S., Velazquez, O., Alejandro, R., Ricordi, C. Combined Treatment of intrapancreatic autologous bone marrow stem cells and hyperbaric oxygen in type 2 diabetes mellitus. *Cell Transplantation*, 2008.

[87] Da Silva Meirelles, L. et al. Mesenchymal stem cells reside in virtually all post-natal organs and tissues. *J. Cell Sci.*, 119:2204–2213, 2006.

[88] Chen, L.B. et al. Differentiation of rat marrow mesenchymal stem cells into pancreatic islet beta-cells. *World J. Gastroenterol.*, 10:3016–3020, 2004.

[89] Choi, K.S. et al. *In vitro* trans-differentiation of rat mesenchymal cells into insulin-producing cells by rat pancreatic extract. *Biochem. Biophys. Res. Commun.*, 330:1299–1305, 2005.

[90] Sun, J. et al. Expression of Pdx-1 in bone marrow mesenchymal stem cells promotes differentiation of islet-like cells *in vitro. Sci. China C Life Sci.*, 49:480–489, 2006.

[91] Li, Y. et al. Generation of insulin-producing cells from PDX-1 gene-modified human mesenchymal stem cells. *J. Cell Physiol.*, 211:36–44, 2007.

[92] Karnieli, O. et al. Generation of insulin-producing cells from human bone marrow mesenchymal stem cells by genetic manipulation. *Stem Cells*, 25:2837–2844, 2007.

[93] Xu, J. et al. Reversal of diabetes in mice by intrahepatic injection of bone-derived GFP-murine mesenchymal stem cells infected with the recombinant retrovirus-carrying human insulin gene. *World J. Surg.*, 31:1872–1882, 2007.

[94] Masaka, T. et al. Derivation of hepato-pancreatic intermediate progenitor cells from a clonal mesenchymal stem cell line of rat bone marrow origin. *Int. J. Mol. Med.*, 22:447–452, 2008.

[95] Moriscot, C. et al. Human bone marrow mesenchymal stem cells can express insulin and key transcription factors of the endocrine pancreas developmental pathway upon genetic and/or microenvironmental manipulation *in vitro. Stem Cells*, 23:594–603, 2005.

[96] Wu, X. H. et al. Reversal of hyperglycemia in diabetic rats by portal vein transplantation of islet-like cells generated from bone marrow mesenchymal stem cells. *World J. Gastroenterol.*, 13:3342–3349, 2007.

[97] Hisanaga, E. et al. A simple method to induce differentiation of murine bone marrow mesenchymal cells to insulin-producing cells using conophylline and betacellulin-delta4. *Endocr. J.*, 55:535–543, 2008.

[98] Chang, C. et al. Mesenchymal stem cells contribute to insulin-producing cells upon microenvironmental manipulation *in vitro. Transplant. Proc.*, 39:3363–3368, 2007.

[99] Baertschiger, R.M. et al. Mesenchymal stem cells derived from human exocrine pancreas express transcription factors implicated in beta-cell development. *Pancreas*, 37:75–84, 2008.

[100] Ballen, K. Challenges in umbilical cord blood stem cell banking for stem cell reviews and reports. *Stem Cell Rev.*, 6:8–14.

[101] Hollands, P. et al. Private cord blood banking: current use and clinical future. *Stem Cell Rev.*, 5:195–203, 2009.

[102] Newcomb, J.D. et al. Umbilical cord blood research: current and future perspectives. *Cell Transplant*, 16:151–158, 2007.

[103] Samuel, G.N. et al. Umbilical cord blood banking: public good or private benefit? *Med. J. Aust.*, 188:533–535, 2008.

[104] Tremolada, C. et al. Adipocyte transplantation and stem cells: Plastic surgery meets regenerative medicine. *Cell Transplant*, 19:1217–1223.

[105] Kang, H.M. et al. Insulin-secreting cells from human eyelid-derived stem cells alleviate type I diabetes in immunocompetent mice. *Stem Cells*, 27:1999–2008, 2009.

[106] Le Douarin, N.M. et al. Neural crest cell plasticity and its limits. *Development*, 131:4637–4650, 2004.

[107] Dong, Q.Y. et al. Allogeneic diabetic mesenchymal stem cells transplantation in streptozotocin-induced diabetic rat. *Clin. Invest. Med.*, 31:E328–337, 2008.

[108] Chang, C. et al. *Mesenchymal Stem Cells Adopt Beta-Cell Fate upon Diabetic Pancreatic Microenvironment.* Pancreas, 2008.

[109] Chang, C.F. et al. Fibronectin and pellet suspension culture promote differentiation of human mesenchymal stem cells into insulin producing cells. *J. Biomed. Mater. Res. A*, 86:1097–1105, 2008.

[110] Ezquer, F.E. et al. Systemic administration of multipotent mesenchymal stromal cells reverts hyperglycemia and prevents nephropathy in type 1 diabetic mice. *Biol. Blood Marrow Transplant*, 14:631–640, 2008.

[111] Paredes, B., Santana, A., Arribas, M., Vicente-Salar, N., de Aza, P., and Roche, E. Phenotypic differences during the osteogenic differentiation of single-cell derived clones isolated from human lipoaspirates. *Journal of Tissue Engineering and Regenerative Medicine*, 2010, in press.

[112] Deutsch, G. et al. A bipotential precursor population for pancreas and liver within the embryonic endoderm. *Development*, 128:871–881, 2001.

[113] Zaret, K.S. Liver specification and early morphogenesis. *Mech. Dev.*, 92:83–88, 2000.

[114] Melton, D. Signals for tissue induction and organ formation in vertebrate embryos. *Harvey Lect.*, 93:49–64, 1997.

[115] Jung, J. et al. Initiation of mammalian liver development from endoderm by fibroblast growth factors. *Science*, 284:1998–2003, 1999.

[116] Lemaigre, F. et al. Liver development update: new embryo models, cell lineage control, and morphogenesis. *Curr. Opin. Genet. Dev.*, 14:582–590, 2004.

[117] Tremblay, K.D. et al. Distinct populations of endoderm cells converge to generate the embryonic liver bud and ventral foregut tissues. *Dev. Biol.*, 280:87–99, 2005.

[118] Yoshitomi, H. et al. Endothelial cell interactions initiate dorsal pancreas development by selectively inducing the transcription factor Ptf1a. *Development*, 131:807–817, 2004.

[119] Zaret, K.S. Hepatocyte differentiation: From the endoderm and beyond. *Curr. Opin. Genet. Dev.*, 11:568–574, 2001.

[120] Wells, J.M. et al. Vertebrate endoderm development. *Annu. Rev. Cell Dev. Biol.*, 15:393–410, 1999.

[121] Rao, M.S. et al. Almost total conversion of pancreas to liver in the adult rat: A reliable model to study transdifferentiation. *Biochem. Biophys. Res. Commun.*, 156:131–136, 1988.

[122] Rao, M.S. et al. Hepatic transdifferentiation in the pancreas. *Semin. Cell Biol.*, 6:151–156, 1995.

[123] Rao, M.S. et al. Induction of hepatocytes in the pancreas of copper-depleted rats following copper repletion. *Cell Differ.*, 18:109–117, 1986.

[124] Shen, C.N. et al. Molecular basis of transdifferentiation of pancreas to liver. *Nat. Cell Biol.*, 2:879–887, 2000.

[125] Wolf, H.K. et al. Exocrine pancreatic tissue in human liver: A metaplastic process? *Am. J. Surg. Pathol.*, 14:590–595, 1990.

[126] Lee, B.C. et al. Metaplastic pancreatic cells in liver tumors induced by diethylnitrosamine. *Exp. Mol. Pathol.*, 50:104–113, 1989.

[127] Jonsson, J. et al. Insulin-promoter-factor 1 is required for pancreas development in mice. *Nature*, 371:606–609, 1994.

[128] Ahlgren, U. et al. beta-cell-specific inactivation of the mouse Ipf1/Pdx1 gene results in loss of the beta-cell phenotype and maturity onset diabetes. *Genes Dev.*, 12:1763–1768, 1998.

[129] Ferber, S. et al. Pancreatic and duodenal homeobox gene 1 induces expression of insulin genes in liver and ameliorates streptozotocin-induced hyperglycemia. *Nat. Med.*, 6:568–572, 2000.

[130] Ber, I. et al. Functional, persistent, and extended liver to pancreas transdifferentiation. *J. Biol. Chem.*, 278:31950–31957, 2003.

[131] Meivar-Levy, I. et al. Pancreatic and duodenal homeobox gene 1 induces hepatic dedifferentiation by suppressing the expression of CCAAT/enhancer-binding protein beta. *Hepatology*, 46:898–905, 2007.

[132] Tang, D. Q. et al. Reprogramming liver-stem WB cells into functional insulin-producing cells by persistent expression of Pdx1- and Pdx1-VP16 mediated by lentiviral vectors. Lab Invest, 86 83–93, 2006.

[133] Wang, A.Y. et al. Adenovirus transduction is required for the correction of diabetes using Pdx-1 or Neurogenin-3 in the liver. *Mol. Ther.*, 15:255–263, 2007.

[134] Kaneto, H. et al. A crucial role of MafA as a novel therapeutic target for diabetes. *J. Biol. Chem.*, 280:15047–15052, 2005.

[135] Kaneto, H. et al. Role of PDX-1 and MafA as a potential therapeutic target for diabetes. *Diabetes Res. Clin. Pract.*, 77, Suppl. 1:S127–137, 2007.

[136] Kaneto, H. et al. Crucial role of PDX-1 in pancreas development, beta-cell differentiation, and induction of surrogate beta-cells. *Curr. Med. Chem.*, 14:1745–1752, 2007.

[137] Matsuoka, T.A. et al. MafA regulates expression of genes important to islet beta-cell function. *Mol. Endocrinol.*, 21:2764–2774, 2007.

[138] Miyatsuka, T. et al. Ectopically expressed PDX-1 in liver initiates endocrine and exocrine pancreas differentiation but causes dysmorphogenesis. *Biochem. Biophys. Res. Commun.*, 310:1017–1025, 2003.

[139] Kojima, H. et al. NeuroD-betacellulin gene therapy induces islet neogenesis in the liver and reverses diabetes in mice. *Nat. Med.*, 9:596–603, 2003.

[140] Zhou, Q. et al. *In vivo* reprogramming of adult pancreatic exocrine cells to beta-cells. *Nature*, 455:627–632, 2008.

[141] Domínguez-Bendala, J., Ricordi, C. and Pastori, R. Protein transduction: A novel approach to induce *in vitro* pancreatic differentiation. *Cell Transplantation*, 15:85–90, 2006.

[142] Domínguez-Bendala, J. et al. TAT-mediated neurogenin 3 protein transduction stimulates pancreatic endocrine differentiation *in vitro*. *Diabetes*, 54:720–726, 2005.

[143] Warren, L. et al. Highly efficient reprogramming to pluripotency and directed differentiation of human cells with synthetic modified mRNA. *Cell Stem Cell*, 7:618–630.

10

Insight into Muscle Regeneration: The Role of Stem Cells, Tissue Niche and IGF-1

Manuela De Rossi[1], Laura Pelosi[1], Laura Barberi[1],
Bianca Maria Scicchitano[1] and Antonio Musarò[1,2]

[1]*Institute Pasteur Cenci-Bolognetti, DAHFMO-Unit of Histology and
Medical Embryology, IIM, University of Rome "La Sapienza", Rome, Italy;
e-mail: antonio.musaro@uniroma1.it*
[2]*Edith Cowan University, Western Australia*

Abstract

The capacity of adult tissues to regenerate in response to injury stimuli represents an important homeostatic process. Regeneration of adult tissues is a highly coordinated program that partially recapitulates the embryonic developmental program.

The repair of the mammalian tissues is fast becoming a feasible goal of regenerative medicine. The appropriate source of cells for these therapeutic applications is hotly debated, but the technical feasibility of using stem cell therapy to aid in the replacement of injured tissues is well within realistic projections. Although resident progenitor cell populations have now been identified in several tissues, including heart, brain, and skeletal muscles, novel approaches are required to overcome the insufficiencies of endogenous stem cells to alleviate acute and chronic damages. One of the crucial parameters of tissue regeneration is the microenvironment in which the stem cell populations should operate. Stem cell microenvironment, or niche, provides essential cues that regulates stem cell proliferation and that directs cell fate decisions and survival.

P. Di Nardo (Ed.), Adult Stem Cell Standardization, 127–133.

It is therefore plausible that loss of control over these cell fate decisions might lead to a pathological transdifferentiation or cellular transformation.

Keywords: stem cells, muscle regeneration, tissue niche, growth factors.

10.1 Introduction

Although the genetic defects responsible for muscle dysfunction in several inherited pathologies have been well characterised, the molecular basis of muscle wasting remains elusive.

It is generally accepted that the primary cause of functional impairment in muscle is a cumulative failure to repair damage related to an overall decrease in anabolic processes. The functional performance of skeletal muscle tissues declines during post-natal life and it is compromised in different diseases, due to an alteration in muscle fiber composition and an overall decrease in muscle integrity as fibrotic invasions replace functional contractile tissue [1]. Characteristics of skeletal muscle aging and diseases include a conspicuous reduction in myofiber plasticity (due to the progressive loss of muscle mass and in particular of the most powerful fast fibers), alteration in muscle-specific transcriptional mechanisms, and muscle atrophy [1, 2]. An early decrease in protein synthetic rates is followed by a later increase in protein degradation, to affect biochemical, physiological, and morphological parameters of muscle fibers during the aging process. Alterations in regenerative pathways also compromise the functionality of muscle tissues.

10.2 The Role of Stem Cells on Muscle Regeneration

Skeletal muscle regeneration is a coordinate process in which several factors are sequentially activated to maintain and preserve muscle structure and function upon injured stimuli.

A major role in growth, remodeling and regeneration is played by satellite cells, a quiescent population of myogenic cells that reside between the basal lamina and plasmalemma and are rapidly activated in response to appropriate stimuli [3]. The activated satellite cells proliferate as indicated by the expression of factors involved in cell cycle progression and by incorporation of BrDU and [3H] thimidine. Ultimately the committed satellite cells fuse with each other or to the existing fibers to form new muscle fibers during regeneration and muscle repair.

More recently, it has been suggested that other "non-muscle" stem cell populations can participate in muscle regeneration and in some way contribute to maintain the pool of satellite cells [3]. These stem cell populations could either reside within muscle, or be recruited via the circulation in response to homing signals emanating from injured skeletal muscle. These populations include endothelial-associated cells, interstitial cells and bone marrow-derived side population cells.

Nevertheless, if skeletal muscle possesses a stem cell compartment it is not clear why the aged muscle fails to regenerate. Either the resident muscle stem cells drastically decrease during aging or perhaps the senescent muscle is a prohibitive environment for stem cell activation and function.

10.3 Role of Tissue Niche and IGF-1 on Muscle Regeneration

Current advances in stem cell biology justify a cautious optimism. Several instances of evidence suggest that while stem cells represent an important determinant for tissue regeneration, a "qualified" environment is necessary to guarantee and achieve functional results [3].

Specific factors are required to trigger stem cells toward a specific lineage, to improve their survival, and to render them effective in contributing to tissue repair. Studies on stem cell niche lead to the identification of critical players and physiological conditions that improve tissue regeneration and repair [3, 4]. Among these, the insulin-like growth factor (IGF-1) has been involved in the regulation of muscle regeneration and homeostasis [2, 5] (Figure 10.1).

We have previously documented the regenerative properties of a locally acting isoform of IGF-1 (mIGF-1) in skeletal muscle and its dramatic promotion of cell survival and renewal in senescent muscle [6]. Expressed as a muscle-specific transgene [6] or on a viral vector [7], mIGF-1 elicits a striking increase in skeletal muscle mass and strength, a rapid restoration of injured muscle, reducing scar formation.

The anabolic effects of mIGF-1 may be due in part to stimulation of activation of satellite cells.

It is not known whether in mIGF-1 transgenic animals, the satellite cells have an increased ability for self-renewal or whether there is an increased recruitment of non-satellite cells.

Our experimental evidence indicates that mIGF-1 promotes the two suggested pathways which can be considered two temporally separated events of the same biological process.

We demonstrated that upon muscle injury, stem cells expressing c-Kit, Sca-1, and CD45 antigens increased locally and the percentage of the recruited cells was conspicuously enhanced by mIGF-1 expression [8]. FACS profiles of tissues from wild type and MLC/mIGF-1 transgenic mice [6], whose muscles were injured with cardiotoxin, revealed a consistent increase of Side Population (SP) cells in the bone marrow, compared to non-injured controls which percentage increased in MLC/mIGF-1 transgenic mice [8]. In contrast, the number of SP cells remained unchanged in other tissues, such as the spleen and the liver of both wild type and MLC/mIGF-1 transgenic mice after muscle injury [8]. Thus, humoral signals emanating from the injured muscles were sufficient to induce stem cell proliferation in the bone marrow, but not in other tissues. Indeed, treatment of injured mice with 5-fluorouracil (5-FU), a cytotoxic agent which depletes cycling stem cells, was sufficient to block proliferation of bone marrow SP and expansion of the CD45+/Sca-1+ population in injured muscle [8].

In addition, to definitively demonstrate the recruitment of bone marrow-stem cells into the site of muscle injury and the role of mIGF-1 in this process, we performed a bone marrow transplantation of both wild type and MLC/mIGF-1 transgenic mice using bone marrow-SP cells of c-kit/GFP transgenic mice [8]. FACS analysis revealed the presence of c-kit/GFP positive cells in the injured area of skeletal muscle and confirmed the effect of mIGF-1 in the enhancement of circulating stem cells into the site of muscle injury [8].

To follow the recruitment of these cells in regenerating muscle tissue, we performed histochemical analysis of wild type and MLC/mIGF-1 transgenic muscle after cardiotoxin injection using antibodies against Sca-1, as well as markers of myogenic commitment and differentiation. Regenerating muscles express Sca-1 and Pax-7, [3] a homeodomain protein implicated in the specification of satellite cells from multipotent stem cells and muscle regeneration.

Increased Sca-1 positive cells in the region of mononuclear infiltration and in blood vessels of MLC/mIGF-1 damaged muscle suggested a mobilization of stem cell population in the injured area [8]. Interestingly, a sub-population of cells adjoining a vessel expressed Pax-7, whereas vascular tissues did not express Pax-7, suggesting that vessel is the niche of circulating stem cells which, penetrating the interstitium, receive myogenic signals and are committed to the muscle phenotype, participating therefore in muscle regeneration and repair.

One of the critical components of muscle regeneration is the inflammatory response. Indeed, inhibition of the inflammatory response or alteration in the phagocytic activity of myeloid cells impairs the regenerative process [9–12]. Mouse strains with slower rates of phagocytic removal of muscle debris show slower rates of muscle regeneration [13]. These correlations support the expectation that phagocytosis is a necessary feature of muscle repair. However, the inflammatory response must be resolved to proceed toward the final steps of muscle regeneration. In fact, functional impairment is associated with perturbed spatial distribution of inflammatory cells, altered identity of the inflammatory infiltrate (cell type and magnitude of influx) and disrupted temporal sequence, resulting in a persistent rather than resolved inflammatory phase [14].

In a recent work, we identified a potential temporal window to modulate the inflammatory response in skeletal muscle subject to injury, induced by cardiotoxin injection [15]. We demonstrated that local expression of mIGF-1 transgene accelerates the regenerative process of injured skeletal muscle, modulating the inflammatory response and limiting fibrosis [15]. At the molecular level, mIGF-1 expression significantly down-regulated proinflammatory cytokines, such as tumor necrosis factor (TNF)-alpha and interleukin (IL)-1beta, and modulated the expression of CC chemokines involved in the recruitment of monocytes/macrophages [15]. Analysis of the underlying molecular mechanisms revealed that mIGF-1 expression modulated key players of inflammatory response, such as macrophage migration inhibitory factor (MIF), high mobility group protein-1 (HMGB1), and transcription NF-kB [15]. The rapid restoration of injured mIGF-1 transgenic muscle was also associated with connective tissue remodelling and a rapid recovery of functional properties. By modulating the inflammatory response and reducing fibrosis, supplemental mIGF-1 creates a qualitatively different environment for sustaining more efficient muscle regeneration and repair (Figure 10.1).

10.4 Conclusions

These results establish mIGF-1 as a potent enhancer of stem cell-mediated regeneration and provide a baseline to develop strategies to improve muscle regeneration in muscle diseases (Figure 10.1).

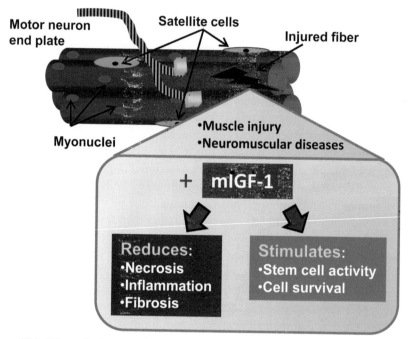

Figure 10.1 Schematic diagram of potential role of mIGF-1 on muscle regeneration. Muscle regeneration is compromised in several pathological conditions probably due to the presence of an hostile microenvironment that alters the activity of stem cells. mIGF-1 promotes a "qualified" environment, by the reduction of necrosis, inflammation and fibrosis, that guarantees a functional activation of stem cells.

In addition, several pieces of evidence suggest that while stem cells represent an important determinant for tissue regeneration, a "qualified" environment is necessary to guarantee and achieve functional results.

In this context, therapeutic applications of adult stem cells to aged or pathological tissue repair in the context of regenerative medicine will require an increased understanding of stem-cell biology, the environment of the aged/pathological tissue and the interaction between the two.

Acknowledgements

Work in the authors' laboratory has been supported by Seventh Framework Programme-Myoage (Grant Agreement Number 223576), Telethon, Fondazione Roma, AFM. Bianca Maria Scicchitano is supported by ASI.

References

[1] Vinciguerra M., Musaro A., Rosenthal N. Regulation of muscle atrophy in aging and disease. *Adv. Exp. Med. Biol.*, 694:211–233, 2010.

[2] Scicchitano, B.M., Rizzuto, E., Musarò, A. Counteracting muscle wasting in aging and neuromuscular diseases: The critical role of IGF-1. *Aging*, 5:1–7, 2009.

[3] Carosio S., Berardinelli M.G., Aucello M., Musarò A. Impact of ageing on muscle cell regeneration. *Ageing Res. Rev.*, 10:35–42, 2011.

[4] Gopinath S.D., Rando T.A. Stem cell review series: aging of the skeletal muscle stem cell niche. *Aging Cell.*, 7:590-598, 2008.

[5] Musarò, A., Rosenthal, N. The critical role of Insulin-like Growth Factor-1 isoforms in the physiopathology of skeletal muscle. *Curr. Genomics*, 3:19–32, 2006.

[6] Musarò, A., McCullagh, K., Paul, A., Houghton, L., Dobrowolny, G., Molinaro, M., Barton, E.R., Sweeney, H.L., Rosenthal, N. Localized Igf-1 transgene expression sustains hypertrophy and regeneration in senescent skeletal muscle. *Nat. Genet.*, 27:195–200, 2001.

[7] Barton-Davis, E.R., Shoturma, D.I., Musarò, A., Rosenthal, N., Sweeney, H.L. Viral mediated expression of insulin-like growth factor I blocks the aging-related loss of skeletal muscle function. *Proc. Natl. Acad. Sci. USA*, 95:15603–15607, 1998.

[8] Musarò A., Giacinti C., Borsellino G., Dobrowolny G., Pelosi L., Cairns L., Ottolenghi S., Bernardi G., Cossu G., Battistini L., Molinaro M., Rosenthal N. Muscle restricted expression of mIGF-1 enhances the recruitment of stem cells during muscle regeneration. *Proc. Natl. Acad. Sci. USA*, 101:1206–1210, 2004.

[9] Grounds M.D. Phagocytosis of necrotic muscle in muscle isografts is influenced by the strain, age, and sex of host mice. *J. Pathol.* 153:71–82, 1987.

[10] Tidball J.G., Wehling-Henricks, M. Macrophages promote muscle membrane repair and muscle fibre growth and regeneration during modified muscle loading in mice in vivo. *J. Physiol.*, 578:327–336, 2007.

[11] Summan M., Warren G.L., Mercer R.R., Chapman R., Hulderman T., Van Rooijen N., Simeonova P.P. Macrophages and skeletal muscle regeneration: A clodronate-containing liposome depletion study. *Am. J. Physiol. Regul. Integr. Comp. Physiol.*, 290:R1488–1495, 2006.

[12] Tidball J.G. Inflammatory processes in muscle injury and repair. *Am. J. Physiol. Regul. Integr. Comp. Physiol.*, 288:R345–R353, 2005.

[13] Teixeira C.F.P., Zamuner S.R., Zuliani J.P., Fernandes C.M., Cruz-Hofling M.A., Fernandes I., Chaves F., Gutierrez J.M. Neutrophils do not contribute to local tissue damage, but play a key role in skeletal muscle regeneration, in mice injected with Bothrops asper snake venom. *Muscle Nerve*, 28:449–459, 2003.

[14] Tidball J.G., Villalta S.A. Regulatory interactions between muscle and the immune system during muscle regeneration. *Am. J. Physiol. Regul. Integr. Comp. Physiol.*, 298:R1173–1187, 2010.

[15] Pelosi L., Giacinti C., Nardis C., Borsellino G., Rizzuto E., Nicoletti C., Wannenes F., Battistini L., Rosenthal N., Molinaro M., Musarò A. Local expression of IGF-1 accelerates muscle regeneration by rapidly modulating inflammatory cytokines and chemokines. *FASEB J.*, 21:1393–1402, 2007.

11

Dynamic Cell Culture in Cardiac Tissue Engineering and Regenerative Medicine

M. Govoni, E. Giordano, C. Muscari, C.M. Caldarera
and C. Guarnieri

*Department of Biochemistry "G. Moruzzi" and National Institute for
Cardiovascular Research (INRC), Università di Bologna, Bologna, Italy;
e-mail: carlo.guarnieri@unibo.it*

Abstract

Regenerating or engineering new cardiac tissue is expected as the replacement therapy for a damaged or failing heart.

Engineering a pseudotissue *in vitro* requires suitable cells, a scaffold – to provide a structured environment with tissue-specific mechanical properties and the ability to integrate with surrounding tissues, and bioreactors – to guarantee environmental conditions and biophysical parameters able to induce, sustain and enhance the development of engineered grafts.

Indeed, bioreactors are essential devices because they provide a dynamic *in vitro* culture environment mimicking *in vivo* conditions for tissue growth. In addition they are also useful for systematic studies of cellular proliferation and differentiation in response to biochemical and physical cues.

The complexity of cardiac tissue easily explains the difficulty to obtain an engineered construct useful for clinical applications and why progress in myocardium regeneration is far from being completed. In addition to optimizing culturing conditions to tackle this ambitious task, the approach based on using a bioreactor also offer the added value to set standards for tools, protocols and physical triggers of cell differentiations to be used in different laboratories interested in this field.

P. Di Nardo (Ed.), Adult Stem Cell Standardization, 135–147.

This short review provides a schematic description of the state of the art in this area.

Keywords: dynamic culture, bioreactors, cardiac tissue engineering, regenerative medicine.

11.1 Introduction

Organ transplantation is the present gold-standard approach to effective treatment of end-stage heart failure following acute myocardial infarction (AMI) [1]. However, several open issues – such as a demanding surgical procedure, the limited number of organs available, organ rejection, and the life-long requirement for immunosuppression therapy – ask for alternative strategies [2]. Regenerating or engineering new cardiac tissue might represent the expected solution for the replacement of a damaged or failing heart.

Emulation of some aspects of normal tissue development and remodelling is the key to future success of tissue engineering, an emerging interdisciplinary field in regenerative medicine. Tissue engineering holds promise for the restoration of tissues and organs damaged by disease, and combines the disciplines of cell and developmental biology to obtain the design and maturation of several tissues [3]. Progress in engineering and regeneration of tissues such as skin and bone has already reached amazing goals. Cardiac muscle tissue engineering is among the present challenges in biomedical research, and several published studies witness the constant advancement in this field.

A number of research groups demonstrated that "cell therapy", i.e. direct injection of cell suspensions in damaged cardiac areas, can help to rescue a damaged heart [4]. In these studies several cell types were tested, such as foetal and neonatal cardiomyocytes, skeletal myoblasts, bone-marrow derived cells, and embryonic and induced pluripotent stem cells [5, 6]. However, the advancement in cell therapy is hampered by the poor survival of injected cells: in fact, most of them die within a few days after grafting into the injured heart. Hypoxia, nutrient deprivation, loss of survival signals and the inflammatory environment where they "land" in a post-AMI tissue are all responsible to contribute to their death [7].

Thus, engineering a pseudotissue *in vitro* was proposed as a more suitable approach than the direct cell injection. In this case, a scaffold must provide a structured environment with tissue-specific mechanical properties and the ability to integrate with surrounding tissues. Biomaterials used as tissue engineering scaffolds include biologically derived supports (e.g. fibrin, hyaluronic acid, collagen, acellular dermis, small intestinal submucosa, alginate, chitin) and biomimetic synthetic polymers (e.g. polyurethane, polylactic and polyglycolic acids and their copolymers, poly(ε-caprolactone), polycarbonate, polyethylene glycol and polyaldehyde hydrogels).

Obviously, the choice of these materials is influenced by the features of the tissue to be engineered. It is therefore required that cardiac muscle engineering might profit of new elastic and deformable scaffolds, adequate for integration into a soft tissue. Moreover, a 3D culture environment should be privileged. A pseudotissue may develop *ex vivo*, involving cell expansion and differentiation in culture and further maturation after implantation, or *in vivo*, where the scaffold is expected to recruit cells from the local environment [8]. In the first option, an original approach is to cultivate the cell/scaffold constructs in a bioreactor, to guarantee environmental conditions and biophysical parameters able to induce, sustain and enhance the development of engineered grafts.

Bioreactors are typical devices initially conceived and used in the field of bacteria fermentation, and then employed in tissue engineering to promote cell and tissue growth.

Their introduction in the field of tissue engineering is justified by the need to apply defined culture protocols [9]. Indeed, bioreactors are essential devices because they provide a dynamic *in vitro* culture environment mimicking *in vivo* conditions for tissue growth. In addition they are also useful for systematic studies of cellular proliferation and differentiation in response to biochemical and physical cues.

In a general way, tissue engineering bioreactors share some common features, such as: establishing a uniform distribution of cells in a 3D scaffold; maintaining the desired concentration of gases and nutrients in the culture medium; exposing developing tissue to physical stimuli.

Functional requirements of the tissue to be engineered determine the specific design requirements of a bioreactor [10]. A specific feature of the cardiac muscle is the coordinated electromechanical coupling among its cells. However, the high degree of complexity of cardiac tissue easily explains the difficulty to obtain an engineered construct useful for clinical applications and why progress in myocardium regeneration is far from being completed. In addition to optimizing culturing conditions to tackle this ambitious task, the approach based on using a bioreactor for engineering a cardiac pseudotissue also offer the added value to set standards for tools, protocols and physical triggers of cell differentiations to be used in different laboratories interested in this field.

This short review provides a schematic description of the state of the art in this area.

11.2 Dynamic Cell Culture for Cardiac Tissue Engineering: The State of the Art

As previously introduced, cardiac tissue engineering, aiming to find solutions for the intrinsic suboptimal regeneration of an injured myocardium by creating *living* tissue replacements, may constitute a new avenue and hope for the treatment of heart disease.

Engineering a fully functional, contractile and robust cardiac tissue means testing different combinations of cells, scaffolds, and specific stimuli in tailored bioreactors.

In particular, we know that mechanical forces influence growth and shape of the tissues in our body, generating changes in intracellular biochemistry and gene expression [11].

At the end of the twentieth century, Carrier et al. [12] used a bioreactor for cardiac tissue engineering purposes for the first time. This device consisted of a rotating flask conceived to overcome the reduction of transport of oxygen and nutrients, characteristic of static cultures, by the mixing of culture medium.

The utilization of rotating vessel bioreactors, which provided dynamic laminar flow patterns, was shown to improve structural and functional construct properties. Moreover, after 1–2 weeks of culture, engineered cardiac constructs expressed proteins and ultrastructural features characteristic of native cardiac tissue.

Studies conducted by the use of rotating bioreactors continued for few years, but despite the fact it was demonstrated that these devices encouraged cells to agglomerate and minimized the shear stress of these tissues, the mixing of culture medium could generate vortexes and damage the growing tissue.

Therefore, an alternative approach was followed culturing cells seeded on thin elastic substrates subjected to appropriate mechanical stimuli.

Akhyari et al. [13] demonstrated that a cyclical mechanical stretch regimen enhanced the formation of a 3D tissue engineered cardiac graft, in accordance with data reported by Kim et al. [14], Ruwhof et al. [15], and Sadoshima et al. [16], who showed that mechanical stress is a potent stimulus for cell proliferation and hypertrophy, as well as for extracellular matrix formation and organization. Human heart cells (from myocardial biopsies obtained from the right ventricular outflow tract) were seeded on a 3D gelatin scaffold and underwent a mechanical stretch regimen using the Bio-Stretch system (ICCT Technologies) [17]. In spite of the fact that this device shows a great versatility in terms of dynamic parameters available, and allows the application of three different stretch regimens simultaneously, the limit of this instrument is the need to use only one single kind of scaffold. In fact, cells must be exclusively cultured on the specific Gelfoam® sponge (Pharmacia & Upjohn Co., Kalamazoo, MI), that is placed in a 35 mm Petri dish with one end glued to the dish, and the other end attached to a coated steel bar. The Petri dish is placed in front of a magnet, and the movement of the steel bar, and the consequent gelatine stretching, allows to transfer the mechanical load to the cells. Dynamic stimulation of cells/gelatine constructs for 14 days at a frequency of 80 cycles/min with a 20% deformation of the initial length generated a marked increase in cell proliferation, improved spatial cell distribution throughout the gelatine scaffold, and markedly increased the total amount of newly synthesized collagen matrix with fibres aligned in parallel to the axis of stress.

One of the most commonly used bioreactors for application of cyclic strain to 2D cell cultures is the Flexcell® Strain Unit (Flexcell Int., Hillsborough, NC) [18]. This commercial device is controlled by a personal computer and allows the use of vacuum pressure to apply cyclic strain. Cells are cultured on flexible-bottomed culture plates and special inserts (Flexcell Int.) are used to create uniaxial or biaxial deformations of the culture silicon membranes. By this device, a cell population can be seeded onto six-well flexible bottom plates, generating a 2D architecture, and subjected to various stretch regimes.

Gwak et al. [19] investigated whether cyclic mechanical strain promotes cardiomyogenesis of embryonic stem cell (ESCs) seeded on elastic polymer poly(lactide-co-caprolactone) (PLCL). Mechanical load was applied by a custom made bioreactor already used by Kim et al. [20] who demonstrated that cyclic mechanical strain enhanced development and function of engineered smooth muscle tissue via an increase of elastin and collagen production and tissue organization. The scaffolds, placed in a culture chamber, were subjected to cyclic strain by periodical back-and-forth movement of a crank. The cyclic strain unit was maintained in a standard incubator for 2 weeks and scaffolds bearing the cells were stretched at 1 Hz frequency with a strain amplitude of 10% of the original scaffold length. Mechanical load promoted and enhanced cardiac-specific gene expression of ESCs *in vitro*. Moreover, tests *in vivo* have shown a significant increase of the grafting efficiency and cardiomyogenetic potential of implanted cells.

Another customized bioreactor was used by Ghazanfari et al. [21] who reasserted that human mesenchymal stem cells (hMSCs) differentiate into smooth muscle cells (SMCs) when exposed to cyclic stretch. Cells were seeded onto elastic medical grade silicon membranes coated with collagen type I, used as cell culture substrate, and then subjected to mechanical stress. This customized device is capable of operating inside an incubator and imposing a uniform tensile strain on cultured cells with strain range of 0–25% and frequency of 1–3 Hz.

Results have provided evidences indicating that uniaxial stretch is a potent regulator of both MSCs proliferation and differentiation, and actin alignment.

Although this device allows the investigation of how mechanical stimulation is involved in development and differentiation of competent cells, the use of a flat membrane only yields a 2D construct, relegating this model to the field of research.

Moon et al. [22] used a tissue bioreactor system to enhance cellular organization and accelerate tissue maturation and formation *in vitro*. In this study, collagen-based decellularized tissue and a custom made bioreactor were used to support and promote cell and tissue growth and differentiation. Human muscle precursor cells (MPCs) were isolated from surgical biopsies of the *psoas* muscle of healthy volunteers. The scaffold, derived from the porcine bladder, was obtained by the extraction of all cellular elements and then cut into strips sutured at both ends to secure the scaffold in the bioreactor. A custom strain device was used for applying the mechanical stimulation, that consisted of cyclic uniaxial stretch and relaxation. A linear actuator was mounted on a tissue culture container where up to 10 scaffold strips could be placed at the same time. MPCs on board were subjected to stretch and relaxation of 10% of their initial length for a periods ranging from 5 days to 3 weeks. No reference about the value of the force applied is presented in the text. Experiments conducted by the application of mechanical load demonstrated that 1 week of bioreactor preconditioning was sufficient to promote cellular maturation and tissue organization. Moreover, in another set of experiments, cell/scaffold constructs were preconditioned for 1 week in a bioreactor and then implanted onto *latissimus dorsi* muscle of athymic mice to evaluate the contractility of the new tissue after 1–4 weeks.

This study confirms the importance and utility of preconditioning *in vitro*, in a bioreactor, to accelerate the maturation of cell population stimulating contractility of the engineered tissue implanted *in vivo*.

Recently, Candiani et al. [23] showed results of application of mechanical load on murine skeletal muscle cells C2C12 seeded onto a biodegradable electrospun microfibrous polyesterurethane scaffold (DegraPol®) in a customized bioreactor. In this device, four sample strips are housed in a Plexiglas® culture chamber and stimulated by a stepping motor equipped with two drive shafts and four pairs of grips able to hold the specimens at the extremities. The stretching pattern is composed by a unidirectional stretching phase (24 hr of stretching at 0.02 mm/hr, up to 960 μm displacement) and a cyclic stretch phase (three consecutive displacement controlled 5-pulse bursts of 0.5 Hz frequency, 1 mm amplitude separated by 30 sec rest and followed by 28 min rest). The overall deformation of the samples, including unidirectional stretching phase, measured by authors was 6–7% (<10% in accordance with literature). Results demonstrated that cells trained in bioreactor for 10 days are featured by a relevant increase (eight-fold) in heavy chain myosin (MHC) with respect to corresponding static control. Hence, this study once more underlines the importance of using a device able to apply a mechanical stimulation and opens exciting opportunities for muscle tissue engineering applications.

In 2010, an innovative and different concept of mechanical stimulation was presented by Weiss et al. [24] who developed and tested a computer controlled dynamic bioreactor for continuous ultra-slow uniaxial distraction of 3D cultures. The main advantages asserted by the authors include: (1) application of uniaxial ultra-slow stepless distraction and real-time control of the distraction distance with high accuracy; (2) application of tension strain on a 3D cell culture within a standard CO_2-incubator without use of an artificial culture matrix; (3) possibility of histological investigation without loss of distraction; (4) feasibility of molecular analysis on RNA and protein level.

The most innovative approach is featured by the absence of a cell scaffold or matrix substrate. Cells (MSCs), after centrifugation to obtain a cellular pellet, were transferred to each of the 12 β-tricalcium-phosphate (β-TCP)–anchors rigidly fixed at the sample rack of the distractor.

Although the main distinctive feature of this new distraction model may increase our understanding on the regulatory mechanisms of mechanical strain on the metabolism of stem cells, the use of TCP-blocks, known to preferentially address osteogenic differentiation, is not suitable for promoting the growth of cardiac tissue.

Most recently, the MATE mechanical strain device was presented in the literature [25]. This bioreactor is featured by a portable and practical design with a compact frame for insertion into standard incubators. Its culture chamber imitates a six-well plate where a translucent polysulfone lid interfaces with the culture tray and two butterfly fasteners secure these components to the MATE frame. Force and displacement were controlled and measured by six electromagnetic actuators which load each construct in unconfined compression by raising the specimens into contact with the loading posts. Although mechanical stimulation can be applied in repeatable manner and mechanical properties can be periodically evaluated to map functional development, in this work only the effect of dynamic stimulation of hydrogels and bovine cartilage without cells on board were evaluated. Thus, it will be interesting to investigate the real effectiveness of this device when mechanical load is applied to cell/scaffold constructs.

Another class of bioreactor used in cardiac tissue regeneration is composed by perfusion bioreactors where a mechanical load is transferred to the cells by the passage of culture medium routed through the construct with a perfusion loop.

Brown et al. [26], hypothesizing that provision of pulsatile interstitial medium flow to an engineered cardiac patch would result in enhanced tissue assembly, suggested the use of a perfusion bioreactor. This device, capable of providing pulsatile fluid flow and flow rates, demonstrated to improve contractile properties of constructs subjected to a pulsatile interstitial medium flow.

Cardiac patches, obtained by seeding neonatal rat cardiomyocytes onto Ultrafoam® collagen hemostat discs (Davol, Cranston, RI), were cultured for 5 days in a humidified 5% CO_2 incubator and subjected to two different overall flow rates (1.50 mL/min or 0.32 mL/min pump settings) at frequency of 1 Hz. Data reported in this study show that cultivation under pulsatile flow has beneficial effects on contractile properties and promotes cell hypertrophy.

In the same year, Radisic et al. [27] used interstitial medium flow in conjunction with fibrous poly-glycolic acid scaffolds or with porous collagen scaffolds in order to increase the thickness of viable tissue above $\sim$100 μm, and also overcome diffusion oxygen limitations both cell seeding and tissue cultivation.

Electrical stimulation has been shown to promote cell differentiation and coupling, as evidenced by the presence of striations and gap junctions, and resulted in concurrent development of conductive and contractile properties of cardiac constructs [28].

Thus, the technology linked to the development of electric bioreactors continue to move towards various directions and the final goal of these studies will be to better assess the standards of electrical stimulation with respect to electrode geometry, material properties and charge-transfer characteristics at the electrode-electrolyte interface.

Recently, Serena et al. [29] using a custom-built electrical stimulation bioreactor reported cardiogenesis in embryoid bodies (EBs) derived from human embryonic stem cells. At day 4 and 8 a single electrical field pulse of 1 V/mm and duration of either 1 or 90 s was applied to the cells and their cardiac differentiation was verified by the detection of cardiac troponin T, the presence of sarcomeric structure and spontaneous contractions.

Ultimately, because of the complexity of myocardium tissue, some research groups proposed to apply to the cells a combination of stimuli coupling different features of existing bioreactors.

Barash et al. [30], for instance, used an electric field stimulation integrated into a perfusion bioreactor with the purpose to produce thick and functional cardiac patches; Feng et al. [31] coupled electric with mechanical stimuli to recreate myocardial conditions; Galie et al. [32] described the design and validation of a bioreactor to apply simultaneous mechanical and fluidic stress to 3D cell-seeded collagen hydrogels.

In recent years, our research group aimed at designing a reliable innovative bioreactor, acting as a stand-alone cell culture incubator, easy to operate and effective in addressing towards the muscle phenotype, via the transfer of a controlled cyclic deformation, mesenchymal stem cells seeded onto a 3D bioreabsorbable scaffold. This approach resulted in the submission of an international application (WO/2011/013067) to patent our device. Preliminary results of this work are published in [33]. Briefly, the cells "trained" over a week displayed multilayer organization and invaded the 3D mesh of the scaffold, without the need of any chemical or genetic manipulation. The biochemical and immunohistochemical analysis of the pseudotissue constructs showed typical markers of muscle cells. This device is thus proposed as new basic research tool in the field of tissue engineering and for the good manufacturing procedures to obtain pseudotissue constructs to test in regenerative medicine protocols for cardiac repair.

11.3 Conclusions

Tissue engineering and regenerative medicine are rapidly advancing fields that open new and exciting opportunities for revolutionary therapeutic modalities and technologies. The development of viable substitutes that restore the function of damaged tissues and organs is an exciting research perspective. Growing complex pseudotissues is a however a difficult task, that requires

Table 11.1 Bioreactors for cardiac tissue engineering.

Authors	Year	Cell source	Scaffold	Biophysical stimulus
Carrier *et al.*	1999	Cardiac myocytes (neonatal rat and embryonic chick)	PGA	Perfusion by medium mixing
Kim *et al*	1999	SMCs (rat aorta)	PGA and collagen I sponge	Mechanical stretch
Kim *et al*	2000	SMCs (rat aorta)	PGA/PLLA	Mechanical stretch
Akhyari *et al.*	2002	Human cardiac myocytes	Gelfoam ® gelatine	Mechanical stretch
Feng *et al.*	2005	Neonatal cardiac myocytes (rat)	Silicon membranes	Electric field – mechanical stretch
Cevallos *et al.*	2006	HUVECs	Silicon membranes	Mechanical stretch
Gwack *et al.*	2008	ESCs-derived cardiac myocytes	PLCL/PLGA	Mechanical stretch
Moon *et al.*	2008	MPCs (human)	Acellular collagen matrices	Mechanical stretch
Brown *et al*	2008	Neonatal cardiac myocytes (rat)	Ultrafoam ® collagen	Perfusion
Radisic *et al*	2008	Neonatal cardiac myocytes (rat)	PGS and collagen	Perfusion
Tandon *et al.*	2008	Neonatal cardiac myocytes (rat)	None	Electric field
Ghazanfari *et al.*	2009	MSCs (human)	Silicon membrane	Mechanical stretch
Serena *et al.*	2010	ESCs	None	Electric field
Candiani *et al.*	2010	C2C12	DegraPol ®	Mechanical stretch
Weiss *et al.*	2010	MSCs (human)	None	Ultra-slow uniaxial distraction
Barash *et al.*	2010	Neonatal cardiac myocytes (rat)	Alginate	Electric field – perfusion
Lujan *et al.*	2010	None	Hydrogels and bovine cartilage	Mechanical load
Govoni *et al.*	2010	MSCs (rat)	Hyalonect ®	Mechanical stretch
Galie *et al.*	2011	Cardiac fibroblasts (rat)	Collagen hydrogel	Mechanical stretch – fluidic stress

SMCs: Smooth Muscle Cells; HUVECs : Human umbilical Vein Endothelial Cells; ESCs: Embryonic Stem Cells; MSCs: Mesenchymal Stem Cells; MPCs: Muscle Precursor Cells; C2C12: mouse myoblast cell line.

PGA: poly-glycolic acid; PLLA: poly-(L-lactic acid); PLCL: poly(lactide-co-caprolactone); PLGA: poly(lactide-co-glycolide); PGS: poly(glycerol-sebacate) the teamwork of biologists, material scientists, engineers and clinicians. Although *in vitro* tissue engineering has significantly enhanced our understanding of cell and developmental biology, it is now time to move towards the clinical application of these experiments. The growing understanding of the relationship among cell proliferation and differentiation, 3D architecture of cell/scaffold constructs, and gene/protein expression in a functional engineered pseudotissue, addresses the way to follow for the more promising approach. Bioreactors appear essential in the *ex vivo* engineering of cardiac muscle tissue: state of the art devices promote cell proliferation, differentiation and alignment, improve cell metabolic activity and generate a 3D pseudotissue. These encouraging results suggest that further development of dynamic cell culture protocols will grant an effective approach to the therapy of a damaged cardiac muscle and promise to offer added value to set standards for tools, protocols and physical triggers of cell differentiations to be used in different laboratories involved in this field.

References

[1] J. Herreros, J.C. Trainini and J.C. Chachques, Alternatives to heart transplantation: Integration of biology with surgery. *Frontiers in Bioscience (Elite Ed)*, 3:635–647, 2011.

[2] J. Leor and S. Cohen, Myocardial tissue engineering: creating a muscle patch for a wounded heart. *Annals of the New York Academy of Sciences*, 1015:312–319, 2004.

[3] D.E. Ingber and M. Levin, What lies at the interface of regenerative medicine and developmental biology?. *Development*, 134(14):2541–2547, 2007.

[4] R. Tee, Z. Lokmic, W.A. Morrison and R.J. Dilley, Strategies in cardiac tissue engineering. *ANZ Journal of Surgery*, 80(10):683–693, 2010.

[5] M.K. Soonpaa, G.Y. Koh, M.G. Klug and L.J. Field, Formation of nascent intercalated disks between grafted fetal cardiomyocytes and host myocardium. *Science*, 264(5155):98–101, 1994.

[6] S. Durrani, M. Konoplyannikov, M. Ashraf and K.H. Haider, Skeletal myoblasts for cardiac repair. *Regenerative Medicine*, 5(6):919–932, 2010.

[7] M. Zhang, D. Methot, V. Poppa, Y. Fujio, K. Walsh and C.E. Murry, Cardiomyocyte grafting for cardiac repair: Graft cell death and anti-death strategies. *Journal of Molecular and Cellular Cardiology*, 33(5):907–921, 2001.

[8] A. Oneida, Tissue engineering. *Current Opinion in Otolaryngology & Head and Neck Surgery*, 13:233–241, 2005.

[9] H.C. Chen and Y.C. Hu, Bioreactors for tissue engineering. *Biotechnology Letters*, 28(18):1415–1423, 2006.

[10] K. Bilodeau and D. Mantovani, Bioreactors for tissue engineering: Focus on mechanical constraints. A comparative review. *Tissue Engineering*, 12(8):2367–2383, 2006.

[11] D.E. Jaalouk and J. Lammerding, Mechanotransduction gone awry. *Nature Review Molecular Cell Biology*, 10(1):63–73, 2009.

[12] R.L. Carrier, M. Papadaki, M. Rupnick, F.J. Schoen, N. Bursac, R. Langer, L.E. Freed and G. Vunjak-Novakovic, Cardiac tissue engineering: Cell seeding, cultivation parameters, and tissue construct characterization. *Biotechnology and Bioengineering*, 64(5):580–589, 1999.

[13] P. Akhyari, P.W.M. Fedak, R.D. Weisel, T.Y.J. Lee, S. Verma, D.A.G. Mickle and R.K. Li, Mechanical stretch regimen enhances the formation of bioengineered autologous cardiac muscle grafts. *Circulation*, 106:137–142, 2002.

[14] B.S. Kim, J. Nikolovski, J. Bonadio and D.J. Mooney, Cyclic mechanical strain regulates the development of engineered smooth muscle tissue. *Nature Biotechnology*, 17(10):979–983, 1999.

[15] C. Ruwhof and A. van der Laarse, Mechanical stress-induced cardiac hypertrophy: mechanisms and signal transduction pathways. *Cardiovascular Research*, 47(1):23–37, 2000.

[16] J. Sadoshima and S. Izumo, The cellular and molecular response of cardiac myocytes to mechanical stress. *Annual Review of Physiology*, 59:551–571, 1997.

[17] M. Liu, S. Montazeri, T. Jedlovsky, R. van Wert, J. Zhang, R.K. Li and J. Yan, Biostretch, a computerized cell strain apparatus for three-dimensional organotypic cultures. *In Vitro Cellular & Developmental Biology – Animal*, 35(2):87–93, 1999.

[18] M. Cevallos, G.M. Riha, X. Wang, H. Yang, S. Yan, M. Li, H. Chai, Q. Yao and C. Chen, Cyclic strain induces expression of specific smooth muscle cell markers in human endothelial cells. *Differentiation*, 74(9/10):552–561, 2006.

[19] S.J. Gwak, S.H. Bhang, I.K. Kim, S.S. Kim, S.W. Cho, O. Jeon, K.J. Yoo, A.J. Putnam and B.S. Kim, The effect of cyclic strain on embryonic stem cell-derived cardiomyocytes. *Biomaterials*, 29(7):844–856, 2008.

[20] B.S. Kim and D.J. Mooney, Scaffolds for engineering smooth muscle under cyclic mechanical strain conditions. *Journal of Biomechanical Engineering*, 122(3):210-215, 2000.

[21] S. Ghazanfari, M. Tafazzoli-Shadpour and M.A. Shokrgozar, Effects of cyclic stretch on proliferation of mesenchymal stem cells and their differentiation to smooth muscle cells. *Biochemical and Biophysical Research Communications*, 388(3):601–605, 2009.

[22] D.G. Moon, G. Christ, J.D. Stitzel, A. Atala and J.J. Yoo, Cyclic mechanical preconditioning improves engineered muscle contraction. *Tissue Engineering: Part A*, 14(4):473–482, 2008.

[23] G. Candiani, S.A. Riboldi, N. Sadr, S. Lorenzoni, P. Neuenschwander, F.M. Montevecchi and S. Mantero, Cyclic mechanical stimulation favors myosin heavy chain accumulation in engineered skeletal muscle constructs. *Journal of Applied Biomaterials and Biomechanics*, 8(2):68–75, 2010.

[24] S. Weiss, P. Henle, W. Roth, R. Bock, S. Boeuf and W. Richter, Design and characterization of a new bioreactor for continuous ultra-slow uniaxial distraction of a three-dimensional scaffold-free stem cell culture. *Biotechnology Progress*, 2010 [Epub ahead of print].

[25] T.J. Lujan, K.M. Wirtz, C.S. Bahney, S.M. Madey, B. Johnstone and M. Bottlang, A novel bioreactor for the dynamic stimulation and mechanical evaluation of multiple tissue-engineered constructs. *Tissue Engineering Part C Methods*, 2010 [Epub ahead of print].

[26] M.A. Brown, R.K. Iyer and M. Radisic, Pulsatile perfusion bioreactor for cardiac tissue engineering. *Biotechnology Progress*, 24(4):907–920, 2008.

[27] M. Radisic, A. Marsano, R. Maidhof, Y. Wang and G. Vunjak-Novakovic, Cardiac tissue engineering using perfusion bioreactor systems. *Nature Protocols*, 3(4):719–738, 2008.

[28] N. Tandon, A. Marsano, C. Cannizzaro, J. Voldman and G. Vunjak-Novakovic, Design of electrical stimulation bioreactors for cardiac tissue engineering. *Proceedings IEEE Engineering in Medicine and Biology Society 2008*, 3594–3597, 2008.

[29] E. Serena, E. Figallo, N. Tandon, C. Cannizzaro, S. Gerecht, N. Elvassore and G. Vunjak-Novakovic, Electrical stimulation of human embryonic stem cells: Cardiac differentiation and the generation of reactive oxygen species. *Experimental Cell Research*, 315(20):3611–3619, 2009.

[30] Y. Barash, T. Dvir, P. Tandeitnik, E. Ruvinov, H. Guterman and S. Cohen, Electric field stimulation integrated into perfusion bioreactor for cardiac tissue engineering. *Tissue Engineering Part C: Methods*, 16(6):1417–1426, 2010.

[31] Z. Feng, T. Matsumoto, Y. Nomura and T. Nakamura, An electro-tensile bioreactor for 3-D culturing of cardiomyocytes. *Proceedings IEEE Engineering in Medicine and Biology Society*, 24(4):73–79, 2005.

[32] P. Galie and J.P. Stegemann, Simultaneous application of interstitial flow and cyclic mechanical strain to a 3D cell-seeded hydrogel. *Tissue Engineering Part C: Methods*, 2011 [Epub ahead of print].

[33] M. Govoni, F. Lotti, M. Lannocca, C. Muscari, E. Giordano, F. Bonafè, G. Pasquinelli, S. Valente, C. Guarnieri, C.M. Caldarera and S. Cavalcanti. A novel bioreactor to mechanically stress stem cells in culture. In: *Conference Proceedings – Congresso Nazionale di Bioingegneria*, Cappello A., D'Alessio T., Knaflitz M. and Montevecchi F.M. (Eds.), Pàtron Editore, Bologna, 2010.

12

Design of a Biomimetic Niche for Adult Progenitor Cell Selection and Differentiation

Lissy Krishnan

Thrombosis Research Unit, Sree Chitra Tirunal Institute for Medical Sciences and Technology, Trivandrum, India; e-mail: lissykk@sctimst.ac.in

Abstract

Human peripheral blood mononuclear cell (PBMNC) contains a mixture of progenitor cells with potential to differentiate in to a wide range of lineages. Therefore, use of autologous progenitors from blood circulation could be a good treatment option, however; due to the limited numbers available, *in vitro* cell expansion may be crucial. Isolation of lineage committed progenitors for regenerative medicine is required because other contaminating cells may have adverse effect on the tissue of interest, however; obtaining pure population is often difficult due to the absence of appropriate surface markers. Therefore, anchorage dependent circulating progenitors could be allowed to home to cell-specific biomimetic niche on which only the specific cell type may home, survive and differentiate to the desired cell and multiply to get them in larger numbers. In the proof-of-concept presented here, each culture was started from peripheral blood mononuclear cells (PBMNC) on cell specific matrix compositions which consisted of adhesive proteins, growth factors and glycosaminoglcans. Three cell types of interest were endothelial progenitor cells (EPCs), smooth muscle cell progenitors (SMPCs) and neural progenitor cells (NPC). Differentiation of EPC into endothelial cells (EC), SMPC to smooth muscle cells (SMCs) and NPC to neurons were evaluated by their respective morphological characteristics and cell specific immuno cytochemical signatures. The EC that differentiated from EPC expressed vWF &

P. Di Nardo (Ed.), Adult Stem Cell Standardization, 149–166.

endocytosed AcLDL, SMC that differentiated from SMPC expressed SMA and calponin. The neurons that were developed from NPCs expressed an early marker (β-tubulin III) and a terminal marker (MAP-2).

This chapter puts forward an approach of employing cell-specific niche for differentiation of PBMNC fraction of human blood into EC, SMC and neurons. The usefulness of the concept is established to harvest and multiply these progenitors *in vitro* to obtain sufficient number of required cell types which may find application in regenerative medicine.

Keywords: adult stem cell, circulating progenitors, biomimetic niche, stem cell homing, extracellular matrix, differentiation.

12.1 Introduction

Cell transplantation holds remarkable prospective to repair or regenerate the injured tissues. Direct injection or transplantation of cells may effectively restore small areas of damage. However, repair of large ischemic injuries requires a higher number of cells along with scaffolds for their retention at the site. The main concern in regenerative therapy is to select suitable cells that are immunosurvilant, grow and maintain them in large supply, get them differentiated and ensure integration into functional tissue efficiently after transplantation. Currently experimental transplantation is actively being carried out using cells from fetus tissue, pre-differentiated embryonic stem cells, cord blood derived stem cells and adult stem cells. Fetal cells, embryonic stem cells and cord blood stem cells face complex problems due to immuno rejection, risk of transmitting infections, unavailability of sufficient tissues and moreover, ethical and logical issues. In this scenario adult stem cell possess the potential to act as a source of cells for regenerative medicine.

A number of studies in the last few years provided evidence that adult stem cells (ASC) not only generate lineage-fixed progeny, but similar to embryonic cells, can attain a wider differentiation repertoire. This offers numerous advantages of ASC being used for cellular therapies. Many adult tissues contain populations of stem cells that can self-replicate and give rise to daughter cells that undergo an irreversible terminal differentiation.

However, there is a need to standardize the procedure for harvesting the ASC, handling the cells, and to apply them effectively for tissue regeneration. There are major questions yet to be answered which include: (i) from which tissue can adult stem cell be obtained; (ii) is it necessary to purify them to homogeneity; (iii) how many cells are to be transplanted; (iv) can *in vitro* manipulations produce sufficient cell number; (v) are pre-differentiated progenitors better for tissue regeneration than the differentiated tissue specific cells; (vi) can the differentiation *in vitro* be done without phenotype drift; (vii) how to deliver cells to the site of injury; (viii) what should be the time gap between injury and transplantation; (ix) how to ensure that transplanted cells home to the injured site and survive there; (x) what may be the local effects of injury on transplanted cells etc. Currently, several groups are attempting experimental research and further some clinical studies with an objective to answer one or more questions as stated above. Interpretation of experiments that is done by various groups to make a conclusion is often puzzling mainly because the study design is not comparable to one another. Practically, no single study can answer all the problems therefore; there is a need for simultaneous and coordinated efforts to carry out research with a focus to corroborate point by point. Therefore, an urgent need exist for standardization of research strategy before successful translation of the laboratory finding to clinics can be attained. Meanwhile, efforts should be taken to review the developments in the field periodically and tap the current knowledge for standardization of effective translational stem cell procedures.

12.1.1 Adult Stem Cells for Tissue Regeneration

Adult stem cells are found in many tissues including heart, brain, bone marrow, peripheral blood, blood vessels, skeletal muscle, skin, teeth, heart, gut, liver, ovarian epithelium, and testes. This knowledge has encouraged researchers to work out strategies for application of ASC in transplantation. Typically, there is a very small number of ASC in each tissue, and once removed from the body, their capacity to divide is limited, making generation of large quantities of stem cells or differentiated cells a difficult task. Therefore, an important focus may be to find optimum conditions to grow large quantities of ASC and to manipulate them to generate specific cell types so that they can be used to treat injury or disease. Adult stem cells are thought to reside in a specific area of each tissue called a 'stem cell niche' and a similar cell-specific niche may be necessary for the survival of ASC *in vitro* too.

Various approaches for stem cell therapy may be made available depending on the requirement for regeneration or substitution. If the therapy is for *in vivo* regeneration of tissues, stemness of the cell has to be maintained, and the requirement of the niche may be for survival and multiplication without differentiation. If the purpose of transplantation is to give a partial support for tissue regeneration, cells have to be lineage committed and the niche has to emit signal for cell survival, multiplication and differentiation. If differentiated cells are delivered to the site as tissue equivalent such as an engineered functional product, which is another effective option for certain cases, the stem cells are required to be differentiated using signals generated from the niche. Therefore, depending on the need for each application, a specific niche has to be designed and its efficacy may be demonstrated before they are used in clinics.

12.1.2 Strategies for ASC Expansion

Once the differentiation of adult stem cells can be controlled in the laboratory, these cells may become the basis of transplantation-based therapies. The microenvironment in which stem cells reside is termed as 'niche' which regulates stem cell fate. The notion of microenvironments affecting stem cell division and function is not new. Schofield dubbed these 'niches' with respect to hematopoietic stem cells [1] and subsequent reports have described their presence in numerous tissues including germline, neural, skin, intestinal and others [2]. During embryonic development, various niche factors act on embryonic stem cells to alter gene expression, induce their proliferation or differentiation and support survival for the development of the fetus. Within the human body, stem cell niches maintain ASC in a quiescent state. Once there is an injury, the micro-environment actively signals to stem cells leading to either self renewal or differentiation to form new tissues. Characteristics of niche are cell-cell interactions between stem cells, neighboring differentiated cells, adhesion molecules, and the extra cellular matrix (ECM) components. Other stimulants are the oxygen tension, growth factors, cytokines, and physiochemical nature of the environment including the pH, ionic strength and metabolites like adenosine triphosphate. The stem cells and niche may support each other during development and reciprocally signal to maintain each other during adulthood.

Replication of the *in vivo* niche conditions *in vitro* may be useful for regenerative therapies because for successful cell transplantation, proliferation and differentiation must be first controlled in flasks or plates. Human embryonic stem cells are often grown in fibroblastic growth factor-2 containing fetal bovine serum supplemented media. They are grown on a feeder layer of cells, but may not truly mimic *in vivo* niche conditions which causes culture related phenotypic changes in cells. Adult stem cells remain in an undifferentiated state throughout adult life. However, when they are cultured *in vitro*, they often undergo an 'aging' process in which their morphology is changed and their proliferation capacity is decreased. This finding emphasizes the requirement of a biomimetic niche for *in vitro* growth and expansion of stem cells. Depending on the requirement for transplantation, culture conditions of ASC should maintain their stemness, undergo lineage commitment or differentiate. The interplay between stem cells and their niche creates the dynamic system that is necessary for sustaining tissues, and for the ultimate design of stem-cell therapeutics. So the simple location of stem cells is not sufficient to define a niche; it must have both anatomic and functional dimensions [3].

12.1.3 Bone Marrow-Derived Adult Stem Cells (BMDASC)

Adult stem cell possesses the potential to act as a source of cells for autologous regenerative therapy. A number of studies in the last few years provided evidence that BMDASCs generate lineage-fixed progeny, with similarity to embryonic ones, and they can attain a wide differentiation repertoire [4]. Among the different sources of ASC bone marrow derived progenitor cells possess importance when the practical concerns about the isolation of cells and chances of autologous therapy are considered. Progenitors are either formed in bone marrow or in various tissues and are released into blood circulation. The exact mechanism of its formation or their origin is not currently understood. It has been speculated that these are migrants from bone marrow or other organs and they assist wound healing process of the body. There is also evidence which suggests that they originate from CD34+ hematopoeitic stem cells. According to the knowledge to date progenitor cells in the peripheral blood circulate in low numbers, are adherent and fibroblast-like, clonogenic, share most although not all of the surface markers with bone marrow stromal cells. It contains a subpopulation capable of differentiating along and even beyond mesenchymal lineages. Several studies describe existence of subsets of peripheral blood progenitor cells in normal humans that has a capacity for differentiation into fibroblast, osteoblasts, adipocytes, myocytes, neuronal cells, endothelial cells and epithelial cells lineages [5–7]. These peripheral blood progenitor cells are variously named as monocyte derived mesenchymal progenitor (MOMP), mesenchymal precursor cells found in the blood, BMPCs, PB CFU-Fs, CD34-/low hematopoietic stem cell clones with mesenchymal stem cell characteristics or simply circulating progenitor cells. They are variously characterized using one or a combination of markers like CD34, CD133, CD45, VEGF-R, CD31, Tie-2, thrombomodulin, vascular endothelial-cadherin, von Willebrand factor, VEGF; and HSC c-kit etc. [7, 8]. The numerous and complicated phenotypic and gene expression characteristics make the circulating progenitor cells more elusive and it is considered that circulating progenitor cells possess broad spectrum of plasticity [9].

It is well known that progenitor cell homing is carried out by the combined mechanisms of ECM proteins, glycosaminoglycans (GAGs) and growth factors (GFs). Progenitors possess receptors for different classes of adhesion molecules. Intercellular adhesion molecule (ICAM-1) in the Ig family, very late antigen 4 (VLA-4), integrins, L-selectin, and CD44 are examples of receptors that play a role in the adhesion of progenitor cells to bone marrow [10]. Involvement of circulating endothelial progenitor cells (EPCs) in vascular repair has been an important area of study which began in 1997, when Asahara and colleagues isolated putative progenitors [11]. The discovery that a subtype of PBMNCs can differentiate into endothelial cells [12] has led to a new field of cardiovascular science, including some clinical trials of cellular therapy for limb and myocardial ischemia, recently [13]. It is known that EPCs can be isolated from peripheral, umbilical cord, and bone marrow blood [14]. Walter and coworkers demonstrated that circulating endothelial cells can home to denuded parts of the artery after balloon angioplasty [15]. Homing signals for circulating stem or progenitor cells mostly result from local injury and guide them presumably to the target tissue [16]. Thus it is clear that circulating progenitor cells recognize specific matrix components for cell attachment and repair. Furthermore, it has been reported that blood cells contain progenitors that have the potential to differentiate into either endothelial cells (ECs) or smooth muscle cells (SMCs) *in vitro* depending on the composition of the culture matrix and medium [17].

In addition, presence of neuronal progenitors in circulation has been reported by several researchers [18, 19]. Studies have also indicated that the bone marrow derived progenitor cells possess the capacity to differentiate in to neural lineage *in vitro* [20]. Semi-adherent cells derived from bone marrow and maintained under neural stem cell growth medium generate neurospheres with time and express protein markers like nestin or CD133 that identify NSC. Therefore, it is likely that plastic immature cells may be found in the PBMNC fraction which may be differentiated *in vitro* using mechanical and humoral stimuli. The requirement for each cell type may be different and once lineage committed or differentiated cells are produced, it may be used for therapeutic applications.

12.2 Design of a Niche for Adult Stem Cells

The forces driving stem cell differentiation, or maintaining them in a state of suspended dedifferentiation, include secreted and bound messengers or homing signals. Fine tuning of the microenvironment *in vivo* can be a technology that prevents the need to artificially introduce genetic changes in target cells through gene therapy. It is often difficult to define the niche for *in vitro* culture of adult stem cells because the factors present *in vivo* in the microenvironment and the type of signaling available for cell differentiation and tissue homeostasis are not clearly understood. More than the stem cell multiplication, it may be desirable to produce progenitor cell or differentiated cells in large numbers for regeneration of damaged tissue. Therefore, use of ECM components preferably from a human source may be beneficial for application of the lineage-committed cells for clinical use.

Extra cellular matrix molecules play a major role in modulating growth factor interactions with cells and they may be more effective when immobilized within the matrix rather than in soluble form. This approach has been proven effective for influencing the function of SMC and EC growth in arterial tissue engineering [21, 22]. Fibrin binds a number of ECM components (e.g., fibronectin [FN] and gelatin) and growth factors [GFs] (e.g., fibroblast growth factor [FGF] and vascular endothelial growth factor [VEGF]), making it a matrix with cell adhesion and signaling potential [23]. Fibrin is known for its ability to guide stem cells to the site of injury and may be beneficial in the early phases of peripheral nerve injury in rat models by initiating neurite outgrowth [24, 25]. Fibrin which forms a cohesive network of hemostatic plugs and thrombi at the injured blood vessel may be a temporary matrix for initial support of stem/progenitor cell growth and differentiation. Fibrin as a natural scaffold material has sites to bind and retain growth factors that are released based on the cell-demand [26]. Therefore, fibrin is likely to be a more appropriate scaffold for culture of vascular cells.

Fibronectin is inherently present in the cryoprecipitate used for producing fibrin clot [27]. The FN is cross-linked with fibrin by the action of transglutaminase (FXIII) in cryoprecipitate. The role of FN in cell adhesion, migration, signalling for proliferation and survival is well demonstrated. Its ability to bind to fibrin mediated by transglutaminase activity is extremely important in wound repair. Since its discovery in human tissue, HA and its derivatives have been largely studied and applied in the biomedical arena. Hyaluronan exists in free and tissue-bound form and there is a class of HA-binding proteins known as hyaladherens, among which HA binding functions of fibrinogen and collagen are well-known [28]. Various ECM components such as adhesive proteins, GFs, and GAGs can be easily incorporated with fibrin clot through specific interactions of these molecules with FN. Fibronectin has binding sites for collagen, fibrin, heparin, hyaluronic acid and actin and is found in the ECM of many tissues and is also a plasma protein (0.3 g/L) produced by the liver [29]. Therefore, it is possible to use FN contained in exogenous fibrin clot as a delivery vehicle for HA to the injured tissue.

The concept of using a fibrin based matrix for growing differentiated endothelial cell was tested before it was modified and used for growing circulating endothelial progenitor cells (EPC) and differentiating it into endothelial cells (EC). Two types of differentiated cells that were grown on fibrin based matrix were human umbilical vein endothelial cells (HUVEC) and human saphenous vein endothelial cells (HSVEC). The former is normal cells with good proliferation potential. However, it has been reported to be a difficult cell type to grow in ordinary tissue culture polystyrene. Gelatin coated surfaces were reported to be suitable for HUVEC growth *in vitro*. Even with gelatine coated dishes they are reported to drift phenotype when they are passaged in culture. A fibrin composite has been designed that maintains EC culture through several passages without altering its phenotype [30]. In addition, HSVEC isolated from diseased artery was found to grow on the fibrin matrix. It appeared that at the time of harvest dysfunctional HSVEC was more prominent in culture with high expression of prothrombotic molecules von Willebrand factor (vWF) and plasminogen activator inhibitor (PAI) and low expression of antithrombotic molecules (tPA) and endothelial nitric oxide synthetase (eNOS) expression. When the cells were passaged on the fibrin matrix they were found to re-differentiate to express more tPA and eNOS and down regulate vWF and PAI. On gelatin these cells were found to drift into a more thrombogenic phenotype [31]. It was found that addition of more angiogenic and mitogenic growth factors and glycosaminoglycans to the fibrin matrix has promoted deposition of ECM especially the components of basement membrane by EC [32–34]. The advantage of having a fibrin composite matrix for cell growth and maintenance of angiogenic phenotype with ability to deposit elastin and collagen was demonstrated by comparing cell growth on an in-house designed fibrin matrix against growth on commercially available FN coated culture dishes. It was observed that cell proliferation and survival was comparable on FN-coated and fibrin composite coated plates but there was hardly any ECM deposition when FN-coated culture dishes were used. However, FN is essential in the fibrin composite because when FN sites were blocked in fibrin composite using anti FN antibodies, cell growth and survival was poor. All these studies proved that maintenance of functionally active normal phenotype of cells require multiple signals from matrix on which they grow. It was shown that the matrix can be coated on various biomaterial surfaces for improved HUVEC attachment, proliferation and survival [35–38].

12.2.1 Design of a Cell-Specific Niche

Depending on the cell that has to be grown the composition of niche needs to be different. Both angiogenesis and neurogenesis are closely related processes in normal physiology and the niche may share similar properties. To switch from a quiescent to a sprouting phenotype, endothelial cells require angiogenic growth factors, such as basic fibroblast growth factor (bFGF), as well as interactions with ECM molecules [39, 40]. Several GFs have been implicated in embryonic smooth muscle cell differentiation, including transforming growth factors 1, 3, and platelet-derived growth factor BB (PDGF-BB) [41]. In addition to GFs needed for EPC differentiation into EC, neurotrophines like brain derived neurotrophic growth factor (BDNF), nerve growth factor (NGF) etc. play a major role in neurogenesis, neuronal growth, survival, axonal outgrowth and neuroprotection [42–45]. Another important molecule that plays a major role in the neuronal guidance and survival is hyaluronic acid and is present in the developing nervous system. Therefore, to induce neurogenesis in circulating progenitor cells we need a specific niche containing various factors.

12.2.2 Proof of Concept

The concept of using specific matrices for differentiation of circulating progenitors from human PBMNC fraction was published earlier [46, 47]. The major differences of the specific niche and the pattern of cell growth in each case are discussed here with emphasis on how these cells can be applied for regenerative purpose. For all three cell types, fibrin the natural scaffold along with FN, gelatin and growth factors was used. Another common feature is that all culture experiments were done using unselected PBMNC mixture isolated using Histopaqe 1077. While there are common GFs in the niche there are some specific differences which may be responsible for recruitment and incorporation of neural progenitor cells (NPCs), EPCs and SMPCs. Each cell type requires a coordinated sequence of multi-step adhesive and signalling events including chemo attraction, adhesion, transmigration, and finally the differentiation to neuronal, endothelial and smooth muscle cell phenotype.

12.2.3 Engineering of Matrices

Gelatin, gluteraldehyde, Histopaque 1077 and heparin were from Sigma chemicals (USA), M199, MCDB 131 medium, trypsin-EDTA, antibiotics and ascorbic acid were from GIBCO BRL, USA. Vascular endothelial cell growth factor (VEGF) was prepared from bovine hypothalamus according to the method of Maciang et al. [48]. Platelet releasate was used as source of chemokines and was prepared as described earlier [49]. Fetal bovine serum (FBS) was from GIBCO BRL, USA. The complete medium with 20% FBS, 50 IUml-1heparin, 25 μg ml-1 VEGF, 1 μg ml-1ascorbic acid, 0.8 μg ml-1 platelet releasate and 40 μl ml-1 antibiotics was used in various proportions.

Tissue culture polystyrene (TCPS) (NUNC, Rakslide, Denmark) were coated with fibrin composite as described earlier [31] with some modification. For making EPC-matrix cryoprecipitate was added with 50 μg ml^{-1} angiogenic growth factors whereas for making SMPC-matrix 40 μg ml^{-1} platelet growth factors was added. For making fibrin composite matrix for growing neuronal cell, the cryo precipitated fibrinogen concentrate (1 ml) was added with gelatin (0.2%), AGF (150 μg), PGF (1.0 μg) and HA (10 μg). Coated dishes were lyophilized under sterile atmosphere, in a freeze drier (Edwards, Modulyo 4K, UK). In all cases the PBMNC suspended in complete medium was initially seeded in an uncoated 10 cm^2 culture dish and incubated for 1 h in a humidified incubator under 5% CO_2 at 37°C. After 1 h the medium was removed gently and the unattached cells were collected in freshly added complete medium and were added to EPC-matrix, SMPC-matrix or NPC-matrix.

12.2.4 Isolation of PBMNC and Culture

Use of sheep blood was approved for EPC and SMPC isolation by the Institutional Animal Ethics Committee (IAEC) constituted as per the requirement of CPCSEA, Govt. of India. Fifty ml blood was collected from sheep jugular vein into plastic syringes containing 500USP heparin. PBMNCS was isolated by histopaque-1077 density gradient centrifugation. For NPC isolation Institutional ethics committee (IEC) approved collection of 20 ml blood from healthy volunteers. Briefly, red blood cells were settled by centrifugation at 280 g for 1 h, using Hareus Stratos centrifuge (Hareus, Germany). The plasma from the top was discarded; PBMNC with RBC at the interface was collected and diluted with equal volume of M199 to make up the volume to 8 ml, mixed well and was layered over 7 ml histopaque-1077 and centrifuged at 400 g for 30 min at 25°C. The buffy coat containing PBMNC was carefully separated from the plasma-histopaque interface and washed with serum-free M199 by centrifugation at 250 g at 4°C for 10 min. The washed PBMNC were suspended in complete medium MCDB131.

12.2.5 Tracking Differentiation

12.2.5.1 Characterization of EPC

Morphological analysis of the cell in culture was done periodically. The PBMNC culture on the 3rd day of isolation is shown in Figure 12.1A. There are small clusters of cells initially which has expanded by day 7, changed shape and multiplied around the group (Figure 12.1B). By day 12 the cluster size expanded and was ready for splitting (Figure 12.1C). On passaging it was found that cell grew and showed cobble stone morphology (Figure 12.1D). The endothelial cell phenotype was confirmed by analysis of acetylated low density lipoprotein [Ac-LDL] uptake (Figure 12.1E), cellular expression of vWF (Figure 12.1F) by immunochemical method.

12.2.5.2 Characterization of SMPC

Smooth muscle cell progenitors also formed small clusters by day 3 (Figure 12.2A) and by day 7 many cells in the group were found to change shape and elongate with increase in cell number (Figure 12.2B). Between days 14 to 16 the cluster was packed and ready for splitting (Figure 12.2C). Once split, the cells multiplied and formed a typical hill and valley morphology (Figure 12.2D) in the culture and were found to grow through four passages with out any difficulty. On staining for alpha smooth muscle actin (α-SMA) and calponin, cells stained positive for both markers (Figures 12.2E and F).

12.2.5.3 Characterization of NPC

During NPC culture also, initially the PBMNC showed small groups and unlike EPC and SMPC cultures, cells were found slightly elongated in the initial three to four days of seeding (Figure 12.3A). By day 9, there were very long cells around the cluster (Figure 12.3B). On staining with β-tubulin 3 and MAP-2 cells were found positive for both the neuron specific antigens (Figure 12.3C and D). So other than the morphology cell phenotype was also proven for neurons produced from circulating PBMNC. Unlike other types

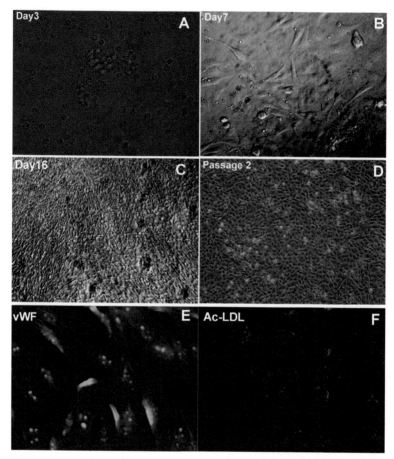

Figure 12.1 Micrographs showing EPC Culture Progress: (A) PBMNC culture on the 3rd day showing cells in cluster; (B) PBMNC culture on the 7th day showing cells with EC-like morphology; (C) proliferated and tightly packed cell cluster-ready to subculture on the 16th day of starting culture; (D) subculture of EC-like cobble-stone morphology seen 3 days after splitting in 1:2 ratio; (E) cells in the 3rd passage characterized as EC by staining with anti-vWF; (F) cells in the 3rd passage demonstrating DilAc-LDL uptake for confirming EC lineage. Magnification of micrographs (A), (B), (E) & (F) is 40×; (C) & (D) is 20×.

the cell multiplication was restricted in the case of NPC; no increase in cell number was found in NPC culture after 6 days.

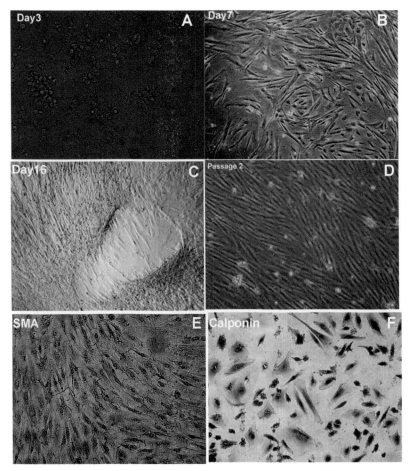

Figure 12.2 Micrographs showing SMPC Culture Progress: (A) PBMNC culture on the 3rd day showing cells in cluster; (B) PBMNC culture on the 7th day showing cells with SMC-like morphology; (C) proliferated and tightly packed cell cluster-ready to subculture on the 16th day of starting culture; (D) subculture of SMC-like hill & valley morphology seen 3 days after splitting in 1:2 ratio; (E) cells in the 3rd passage characterized as SMC by staining with anti-smooth muscle actin (SMA); (F) cells in the 3rd passage characterized as EC by staining with anti-calponin (specific for SMC). Magnification of all micrographs is 40×.

12.3 Conclusion

A natural scaffold composition was used to get EPC, SMPC and NPC differentiated into EC, SMC and neurons respectively, with characteristic features. Use of specifically designed matrices may be used for homing of required

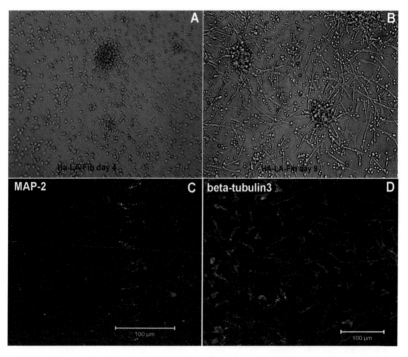

Figure 12.3 Micrographs showing NPC Culture Progress: (A) PBMNC culture on the 3rd day showing cells in cluster; (B) PBMNC culture on the 9th day showing cells with neuron-like morphology; (C) confocal micrograph of MAP-2 (mature neuron marker) stained on the 12th day of starting culture; (D) confocal micrographs of β-tubullin 3 (immature neuron marker) stained on the 9th day of starting culture. Magnification of (A) & (B) is 20×; magnification of (C) & (D) is 40×.

cell type from unselected population of PBMNC isolated from human blood. There are two possible mechanism by which these cells are formed: (1) the progenitor cells found in PBMNC may be plastic and they are induced to differentiate into the specific cells by specific growth factor and GAG combinations, (2) there may be a mixed population of lineage committed progenitors in PBMNC and only the specific type will home to designed matrix composition and survive while the others will senesce because specific component requirement are not met for their survival. More study is required to confirm which of the mechanism is responsible for getting pure population of differentiated EC, SMC or neurons. Many of the components of the matrix in all three types of niche are common such as fibrin, gelatin, FN and certain growth factors. While EC and SMC multiplied and formed large numbers that could be passaged into three to four generations, neuron multiplication was arrested once differentiation occurred. So strategies have to be developed to arrest differentiation of lineage committed cells if they have to multiply *in vitro* or *in vivo* for regeneration of tissues.

Matrix standardization is found to be an important aspect for *in vitro* growth and differentiation of progenitor cells. There may be a similar requirement for *in vivo* survival of progenitors after transplantation. Attention may be given to standardize cell specific niche after identifying the requirements for each type of cell. Such designed matrix composition may turn out to be useful as a delivery vehicle for transplantation of cells for regeneration of damaged tissue.

Acknowledgement

The author wishes to acknowledge the Director, SCTIMST and the Head, BMT Wing for their support and encouragement. The data presented here was earlier published by Ms. Sreerekha and Ms. Anumol Jose.

References

[1] Schofield, R. The relationship between the spleen colony-forming cell and the haemopoietic stem cell. *Blood Cells*, 4:7–25, 1978.

[2] Spradling, A., Drummond-Barbosa, D. and Kai, T. Stem cells find their niche. *Nature*, 414:98–104, 2001.

[3] Scadden D.T. The stem-cell niche as an entity of action. *Nature*, 441:1075–1079, 2006.

[4] Tropepe, V., Hitoshi, S., Sirard, C., Mak, T.W., Rossant, J. and van der Kooy, D. Direct neural fate specification from embryonic stem cells: A primitive mammalian neural stem cell stage acquired through a default mechanism. *Neuron*, 30:65–78, 2001.

[5] Kuwana, M., Okazaki, Y., Kodama, H. Izumi, K. Yasuoka, H., Yoko, O., Kawakami Y. and Ikeda, Y. Human circulating CD14+ monocytes as a source of progenitors that exhibit Mesenchymal cell differentiation. *J. Leukoc. Biol.*, 74:833–845, 2003.

[6] Zhao, Y., Glesne, D. and Huberman, E. A human peripheral blood monocyte-derived subset acts as pluripotent stem cells. *Proc. Natl. Acad. Sci. USA*, 100:2426–2431, 2003.

[7] Kuwana, M., Okazaki, Y., Kodama, H. et al. Human circulating CD14+ monocytes as a source of progenitors that exhibit mesenchymal cell differentiation. *J. Leukoc. Biol.*, 74:833–845, 2003.

[8] Huss, R., Lange, C. and Weissinger, E.M. et al. Evidence of peripheral blood-derived, plastic adherent CD34(-/low) hematopoietic stem cell clones with mesenchymal stem cell characteristics. *Stem Cells*, 18:252–260, 2000.

[9] Porat, Y., Porozov, S., Belkin, D., Shimoni, D., Fisher Y., Belleli, A., Czeiger, D., Silverman, W.F., Belkin, M., Battler, A., Fulga, V. and Savion, N. Isolation of an

adult blood-derived progenitor cell population capable of differentiation into angiogenic, myocardial and neural lineages. *Br. J. Haemat.*, 135:703–714, 2006.

[10] Quesenberry, P.J. and Becker, P.S. Stem cell homing: Rolling, crawling, and nesting. *PNAS*, 26:15155–15157, 1998.

[11] Asahara, T., Murohara, T., Sullivan, A. et al. Isolation of putative progenitor endothelial cells for angiogenesis. *Science*, 275:964–967, 1997.

[12] Szmitko, P.E., Fedak, P.W., Weisel, R.D., Stewart, D.J., Kutryk, M.J. and Verma, S. Endothelial progenitor cells: New hope for a broken heart. *Circulation*, 107:3093–3100, 2003.

[13] Hristov, M., Erl, W. and Weber, P.C. Endothelial progenitor cells: Isolation and characterization. *Trends Cardiovasc. Med.*, 13:201–206, 2003.

[14] Murohara, T. Therapeutic vasculogenesis using human cord blood-derived endothelial progenitors. *Trends Cardiovasc. Med.*, 11:303–307, 2001.

[15] Walter, D.H., Rittig, K., Bahlmann, F.H., Kirchmair, R., Silver, M., Murayama, T., et al. Statin therapy accelerates reendothelialization: A novel effect involving mobilization and incorporation of bone marrow-derived endothelial progenitor cells. *Circulation*, 105:3017–3024, 2002.

[16] Simper, D., Stalboerger, P.G., Panetta, C.J., Wang, S. and Caplice, N.M. Smooth muscle progenitor cells in human. *Blood Circulation*, 106:1199–1209, 2002.

[17] Kaushal, S., Amiel, G.E., Guleserian, K.J., Shapira, O.M., Perry, T., Sutherland, F.W., Rabkin E., Moran, A.M., Schoen, F.J., Atala, A., Soker, S., Bischoff, J., Mayer, J.E. Jr. Functional small diameter neovessels created using endothelial progenitor cells expanded ex vivo. *Nat. Med.* 7:1035–1040, 2001.

[18] Kabos, P., Ehtesham, M., Kabosova, A., Black, K.L. and Yu, J.S. Generation of neural progenitor cells from whole adult bone marrow. *Exp. Neurol.*, 178:288–293, 2002.

[19] Bonilla, S., Silva, A., Valdes, L., Geijo, E., Garcia-Verdugo, J.M., Martinez, S. Functional neural stem cells derived from adult bone marrow. *Neuroscience*, 133:85–95, 2005.

[20] Wislet-Gendebien, S., Wautier, F., Leprince, P., Rogister, B. Astrocytic and neuronal fate of mesenchymal stem cells expressing Nestin. *Brain Research Bulletin*, 68:95–102, 2005.

[21] Mann, B.K., Schmeldlen, R. and West, J.L. Tethered TGF-β increases extra cellular matrix production of vascular smooth muscle cells. *Biomaterials*, 22:439–444, 2001.

[22] Zisch, A.H., Schenk, U., Schense, J.C., Sakiyama-Elbert, S.E. and Hubbell, J.A. Covalently conjugated VEGF-fibrin matrices for endothelialization. *J. Control Release*, 72:101–113, 2001.

[23] Clark, R.A. Fibrin glue for wound repair: facts and fancy. *Thromb. Haemost.*, 90:1003–1006, 2003.

[24] Jua, Yo-El, Janmeya, P.A., McCormickb, M.E., Sawyerc, E.S. and Flanagana, L.A. Enhanced neurite growth from mammalian neurons in three-dimensional salmon fibrin gels. *Biomaterials*, 2097–2108, 2007.

[25] Zeng, L., Worseg, A., Redl, H. and Schlag, G. Fibrin sealant matrix supports outgrowth of peripheral sensory axons. *Scandinavian Journal of Plastic and Reconstructive Surgery and Hand Surgery*, 29:199–204, 1995.

[26] Ehrbar, M., Djonov, V.G., Schnell, C., Tschanz, S.A., Martiny-Baron, G., Schenk, U., Wood, J., Burri, P.H., Hubbell, J.A. and Zisch, A.H. Cell-demanded release of VEGF121

from fibrin implants induces local and controlled blood vessel growth. *Circ. Res.*, 94:1124–1132, 2004.

[27] Perttila, J., Salo, M. and Peltola, O. Plasma fibronectin concentrations in blood products. *Intensive Care Med.*, 16:41–43, 1990.

[28] Leboeuf, R.D., Raja, R.H., Fuller, G.M. and Weoge, P.H. Human fibrinogen specifically binds hyaluronic acid. *J. Biol. Chem.*, 261:12586–12592, 1986.

[29] Millar, T. *Complex Sugars and Their Functions in Biochemistry Explained: A Practical Guide to Learning Biochemistry.* CRC Press, 2002, p. 70.

[30] Prasad, C.K. and Krishnan, L.K. Effect of passage number and matrix characteristics on differentiation of endothelial cells cultured for tissue engineering. *Biomaterials*, 26:5658–5667, 2005.

[31] Prasad, C.K., Jayakumar, K. and Krishnan, L.K. Phenotype gradation of Human Saphenous Vein Endothelial Cell (HSVEC) from cardiovascular disease subjects. *Endothelium*, 13:341–352, 2006.

[32] Divya, P., Sreerekha, P.R. and Krishnan, L.K. Growth factors up regulate deposition and remodeling of ECM by endothelial cells cultured for tissue engineering applications. *Biomolecular Engineering*, 24:593–602, 2007.

[33] Divya, P. and Krishnan, L.K. Design of fibrin matrix composition to enhance endothelial cell growth and extra cellular matrix deposition for in vitro tissue engineering. *Artificial Organs*, 33:16–25, 2009.

[34] Divya, P. and Krishnan, L.K. Glycosaminoglycans restrained in fibrin matrix improve ECM remodeling by endothelial cells grown for vascular tissue engineering. *Tissue Engineering & Regenerative Medicine*, 3:377–388, 2009.

[35] Prasad, C.K., Muraleedharan, C.V. and Krishnan, L.K. Bio-mimetic composite matrix that promotes endothelial cell growth for modification of biomaterial surface. *J. Biomed. Mater. Res. A*, 80:644–654, 2007.

[36] Prasad, C.K. and Krishnan, L.K. Regulation of endothelial cell phenotype by biomimetic matrix coated on biomaterials for cardiovascular tissue engineering. *Acta Biomaterialia*, 4:182–191, 2008.

[37] Divya, P., Philipose, L.P., Palakkal, M., Krishnan, V.K. and Krishnan, L.K. Development of a fibrin composite coated poly (ε-caprolactone) scaffold for potential vascular tissue engineering applications. *J. Biomed. Mater. Res. Part B*, 87B:570–579, 2008.

[38] Divya, P., Krishnan, V.K. and Krishnan, L.K. Vascular tissue generation in response to signaling molecules integrated with a novel poly(ε-caprolactone)-fibrin hybrid scaffold. *Tissue Engineering & Regenerative Medicine*, 1:389–397, 2007, doi: 10.1002/term.48.

[39] Juliano, R.L. and Haskill, S. Signal transduction from the extracellular matrix. *J. Cell Biol.*, 120:577–585, 1993.

[40] Sahni, A., Sporn, L.A. and Francis, C.W. Potentiation of endothelial cell proliferation by fibrin (ogen)-bound fibroblast growth factor-2. *J. Biol. Chem.*, 274:14936–14941, 1999.

[41] Hellstrom, M., Kaln, M., Lindahl, P., et al. Role of PDGF-B and PDGFR-α in recruitment of vascular smooth muscle cells and pericytes during embryonic blood vessel formation in the mouse. *Development*, 126:3047–3055, 1999.

[42] Lazarovici, P., Marcinkiewicz, C. and Lelkes, P. Cross talk between the cardiovascular and nervous systems: Neurotrophic effects of vascular endothelial growth factor (VEGF) and angiogenic effects of nerve growth factor (NGF)-implications in drug development. *Curr. Pharm. Des.*, 12:2609–2622, 2006.

[43] Gritti, A., Parati, E.A., Cova, L., Frolichsthal, P., Galli, R., Wanke, E., Faravelli, L., Morassutti, D.J., Roisen, F., Nickel, D.D. and Vescovi, A.L. Multipotential stem cells from the adult mouse brain proliferate and self-renew in response to basic fibroblast growth factor. *J. Neurosci.*, 16:1091–1100, 1996.

[44] Erlandsson, A., Enarsson, M. and Forsberg-Nilsson, K. Immature neurons from CNS stem cells proliferate in response to platelet-derived growth factor. *The Journal of Neuroscience*, 21:3483–3491, 2001.

[45] Meszar, Z., Felszeghy, S., Veress, G., Matesz, K., Szekely, G. and Modis L. Hyaluronan accumulates around differentiating neurons in spinal cord of chicken embryos. *Brain Research Bulletin*, 75:414–418, 2008.

[46] Sreerekha, P.R., Divya, P. and Krishnan, L.K. Adult stem cell homing and differentiation in vitro on composite fibrin matrix. *Cell Proliferation*, 36:301–312, 2006.

[47] Anumol, J. and Krishnan, L.K. Effect of matrix composition on differentiation of nestin-positive neural progenitors from circulation into neurons. *J. Neural Eng.*, 7, 2010, doi: 10.1088/1741-2560/7/3/036009.

[48] Maciang, T., Cerundolo, J., Ilsley, S. and Kelly, P.R. An endothelial cell growth factor from bovine hypothalamus: Identification and partial characterization. *Proc. Natl. Acad. Sci. USA*, 76:5674–5678, 1979.

[49] Resmi, K.R. and Krishnan, L.K. Protease action and generation of β-thromboglobulin-like protein on platelet activation. *Thrombosis Res.*, 107:23–29, 2002.

13

Aging of Human Bone-Derived Mesenchymal Stem Cells in Their Niche

Günter Lepperdinger

Extracellular Matrix Research Group, Institute for Biomedical Aging Research, Austrian Academy of Sciences, Innsbruck, Austria; e-mail: guenter.lepperdinger@oeaw.ac.at

Abstract

In parallel as we age, stem cells undergo age-related alterations. Often age-associated disturbances are consequences of dysregulated cellular turnover with prevailing decline in tissue regenerative capacity.

Early in life, most, if not all, uncommitted mesenchymal stem cells (MSC) exhibit multipotential differentiation capacity, and their developmental fate is tightly controlled. Later in life, additional types of mesenchymal precursors (MPC) emerge as stable subpopulation.

Besides the rate of cell death, which is gradually increasing, MSC accumulate various forms of damage, which cannot be compensated by cellular repair. Also due to an age-dependent chronic, systemic inflammation, subpopulations of predetermined MPC become manifest. All in all, only little decline of MSC/MPC numbers is envisaged at advanced age, the number of truly naive, multipotential MSC however steadily declines.

Although many questions about aging of MSC are still unanswered, the here presented findings refer to the corresponding niche being a confounding determining factor for the well-being of this ubiquitously present stem cell type. Due to the fact that the number of MSC therapies steadily rises, it is anticipated that this specific theme will also become increasingly important.

P. Di Nardo (Ed.), Adult Stem Cell Standardization, 167–182.

Keywords: stem cell niche, tissue homeostasis, oxygen, inflammaging, VCAM1.

13.1 Introduction

Stem cells are vitally involved in tissue regeneration and homeostasis in late life. Tissue aging is paralleled by a steady decline of tissue regeneration capacity, and thus a slowly but surely approaching loss in function. Yet failures in tight controls are frequently observed at advancing age, and this arising deficit may eventually lead to accumulation of fat deposits in bone, impaired fracture healing, de-regulated hematopoiesis, and autoimmunity.

Mesenchymal stem cells (MSC), also known as multipotent stromal progenitor cells are tissue-specific (or adult) stem cells, which nest in their niche and only commence proliferation upon stimulation, thereby giving rise to one stem cell (self-renewal) and one progenitor cell, which continues to proliferate, and progeny thereof eventually differentiates to regain functional mesenchymal tissue. In order to insure tissue maintenance, each of the processes has to be tightly regulated.

Apparently there are several fundamental questions in the field, which need to be addressed in order to reach a decision, which potentially is the best stem cell source for subsequent biomedical applications:

- How to select high quality MSC with respect to basic stem cell functions as well as their ability to suppress the activation of immune cells?
- What are valid markers and standardized conditions regarding the in vitro propagation of MSC for tissue engineering and immune therapy?
- Which systemic factors impinge on the respective differentiation capacity of mesenchymal progenitors in the context of inflammatory pathologies and necrosis?

These problems share one commonality which actually relates to the *in vivo* niche: specification and maintenance of stemness take place in distinct sites within the body. Meanwhile, relatively little is known about the MSC niche, and even less understood are the instructive measures of the niche, which assure life-long stemness of MSC.

13.1.1 An Appropriate Term: 'Mesenchymal Stem Cell'

Although the designation 'Mesenchymal Stem Cell' is commonly used in the literature, there are many good reasons, which have been scholarly disputed, arguing that this particular appellation is misleading and/or ill-defined; in particular because the definitions for a mesenchymal stem cell are sketchy.

There is no dispute that most blood cell types emerge from meso-dermal derivatives, and hence, hematopoietic cells are of mesenchymal origin. Clearly without ambiguity, 'mesenchymal' is exclusively assigned to non-hematopoietic cells. With respect to the predication 'stem cell', the most stringent condition is a distinct long-term potential to self renew, to-gether with the potency of giving rise to progenitor cells, which in due course differentiate further on to form one or more specialized somatic cell types. Presently, MSC are being isolated and cultured following different methods and protocols, and common characteristics of MSC could only be insufficiently specified, first and foremost because the cell isolates are het-erogeneous and comprise of more than one mesenchymal cell type. It is therefore difficult to resolve how results regarding *ex vivo* propagated cells relate to each other. A large body of information is available on many aspects of cultivated MSC properties. They firmly adhere to cell culture plastic, which is a distinctive functional criterion for selection during isolation. They exhibit clonogenic growth which is a further distinctive selection criterion as during expansion after low density seeding.

Notably, unique and highly specialized cells within the heterogeneous population of mesenchymal stromal cells show, when cultured under spe-cific conditions, a broad range of differentiation potentials. Taken together however, doubts about the appropriateness of the term 'mesenchymal *stem* cell' have been raised, in particular because it is scientifically inaccurate and potentially misleading. Instead, the tentative use of the more appropriate term 'multipotential mesenchymal stromal cell', which is similarly abbreviated by the acronym 'MSC' has been recommended previously [1].

To date it is generally accepted that stromal cells provide the housing for '*mesenchymal progenitor cells*' and hematopoietic stem cells together with theirits respective progeny as well as endothelial progenitor cells. There is reason to believe though that mesenchymal stroma also contains a rare population of naive cell types, which are true mesenchymal stem cells. Most work in this context has focused on isolated MSC in culture. Only recently, it could be demonstrated that MSC can be identified *in vivo* using nestin expression as a distinguishing marker: nestin[+] MSC contain all the bone-marrow colony-forming-unit fibroblastic activity and nestin[+] cells can be propagated as non-adherent 'mesenspheres' that can self-renew and expand in serial transplantations. Also, nestin[+] cells appear to represent bona fide HSC niche cells, and in putting forward a unifying concept for a balanced regulation of haematopoietic and mesenchymal lineages at the stem-cell level, this data suggest that *in vivo* HSC maintenance and MSC proliferation and differentiation are regulated in tandem [2].

13.1.2 Tissue Regeneration and Homeostasis in Late Life

Aging is one, if not the single most important risk factor for degenerative diseases and cancer. Often age-associated disturbances are consequences of dysregulated cellular turnover. It is generally believed that stem cells are vitally involved in tissue regeneration and homeostasis in late life, and thus the question of whether stem cells decline in function is an important issue. For proper function, it is crucial for a stem cell that nests in a quiescent state within its niche to re-enter cell cycle, and as a next step to self-renew, and thereafter become dormant again. The outcome of this process is progeny that is being amplified and giving rise to cells, which in a final step differentiate to regain functional mesenchymal tissue. In order to insure tissue maintenance every single step of the processes has to be tightly regulated.

The loss of regeneration capacity is characteristic for aged tissue. Although the function of traumatized or damaged bone is under normal circumstances restored by mechanically competent bone, osseous defects within aged bone, such as after tooth extraction, or after tumor ablation are barely capable of self-regeneration. This is thought to be due to diminished circulation and perfusion, lacking biomechanical stimuli or fibrosis arising due to excessive cell death and uncontrolled inflammation (Figure 13.1).

13.1.3 Therpeutic Application in Lieu of Rock-Solid Standards

There is growing interest in transplantation of *ex vivo* amplified cell preparations for therapeutic applications. This has been fueled by novel insights from stem cell biology, new molecular tools and promising preclinical model systems, and furthermore due to the fact that MSC can be isolated from most tissues of the human body including bone marrow and adipose tissue [3]. Hence, MSC are considered highly effective therapeutic assets to combat a wide spectrum of diseases: it could be previously shown that culture-expanded MSC exert immune modulatory activities inasmuch as graft-versus-host disease can be ablated [4, 5], as well as based on their wide differentiation potential, defined approaches under the premise of regenerative medicine and tissue engineering with the aim of rebuilding damaged or

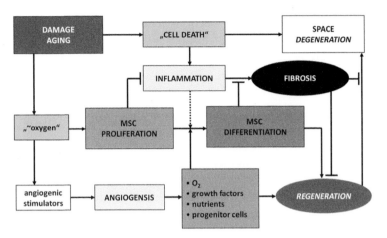

Figure 13.1 Aging and regeneration of organ and tissues. Damage either being introduced through traumatic insults, or accumulating throughout life time, is accompanied by destruction of cells. Eventually these malicious circum-stances yield cell-free spaces within highly structured tissues. These events are accompanied with spreading centers of inflammation. Any anatomical structure blemished in this way is thought to further degenerate without being regenerated by appropriate means. Collaterally to destruction of tissues, break-up of vessels is experienced, which only when countervailed by regulated production of angiogenic stimulators, development of new blood vessels is efficiently initiated. These newly formed structures are in due course facilitating regeneration by transporting essential biological factors and entities to those sites in which regeneration has to be propagated. Center pieces and fundamental pillars for regeneration are tissue-borne stem cells, which when properly invoked counteract foci of inflammation, thus ceasing deployment of fibrotic sites within degenerating tissue. In subsequence amplified progeny incited through bioactive factors which are being delivered through newly formed vasculature commence differentiation into regenerating entities, which adjust for derogating deteriorations and/or clearing age-associated deformation.

diseased organs and body parts, has by now yielded a virtually endless list of only recently initiated clinical trials (see www.trial.gov).

Apparently, the use of MSC as cellular therapeutics demands standardized procedures of isolation and reliable quality control of cell preparations. This, however, is greatly hampered by a multitude of different methods to prepare MSC. Furthermore, there is a growing perception that even under highly standardized culture conditions, continuous effects during long-term culture and eventually replicative senescence need to be taken into account [6–8].

Recently we and others reported that global gene expression profiles change upon long-term culture as well as along with aging *in vivo* [8–

11]. Replicative senescence has been studied in great detail in other cell systems, yet little is known about distinct molecular mechanisms valid for MSC. Apparently application of senescent MSC would exert unforeseeable consequences for cellular therapy.

13.2 Aging of Mesenchymal Stem Cells and Its Niche

It could be clearly demonstrated through work by several groups and by us that MSC face extrinsic and intrinsic aging [12–15]. There are many circumscriptions for aging taking place in a biological system. From the perspective of regenerative processes a valid definition certainly is that the sum of primary restrictions in replenishing multicellular organisms with renewed cellular material is gradually declining [16]. A progressive and irreversible accumulation of DNA damage, which is triggered by telomere shortening yet also other stressors such as oxidative stress, may eventually lead to cellular senescence, which in due course also contributes to organismal aging [17, 18], but how aging is related to the disruption of the homeostatic balance of cell differentiation from a common progenitor is not well understood, as is the capacity of MSC at old age to effect tissue and organ regeneration.

13.2.1 MSC Numbers Marginally Decline at Advanced Age

There are conflicting results regarding the question whether MSC numbers change during life span. MSC can be easily selected within low-density mononuclear cell isolates based on their characteristic of tightly adhering to the plastic surface of a culture dish. Due to their inherent clonogenic growth properties, fibroblastic colonies are formed after extended periods of cultivation and the respective number, also called colony-forming unit-fibroblasts (CFU-f) can be reliably estimated. Yet some laboratories report that total CFU-f decrease with age [7, 19–27] while we and others find only a minor or no significant decline [28–33].

One particular problem with this approach is that there is no agreement on a single, standardized protocol for MSC isolation as well as how to proceed with further analysis of the primary cell isolates. Another issue is that the primary cell isolates are, though at a varying extent, contaminated by other cells, in particular by macrophages and hematopoietic cells. Lastly, single clones within the heterogenous cultures exhibit greatly varying differentiation potential, which was demonstrated by in vitro cell cloning of human stromal cultures: only around 30% of the CFU-f are multipotential and can thus be considered true MSC [34]. In addition to the aforementioned observations, it has also been reported that the number of bi- or tri-potential colonies declines with age [35].

13.2.2 Long-Term Proliferation Potential Ceases with Advancing Age

The proliferative potential of MSC has been extensively investigated. When regarding donor age as the confounding parameter, the published data are less conflicting, because most authors report declining performances in long-term culture [7, 10, 36, 37]. Much of the published data are regarding MSC, which were isolated from bone and bone marrow. Most conspicuously, there are many other sites containing mesenchymal progenitor cells such as the muscle [38], vessel-associated pericytes [39], or blood [40], and evidently the afore-mentioned questions also apply here: as of now, no stringent conclusions can be drawn at this point as most questions in this particular context are still unanswered.

13.2.3 MSC Self-Renewal and Differentiation Potential Is Regulated by Oxygen

When selecting MSC from marrow or osseous surface of cancellous bone, we found them enriched in the endosseous layer. This led to the assumption that MSC biology is regulated by the availability of basic factors such as oxygen. In due course we could show that their differentiation potential is greatly dependent on the availability of oxygen (Figure 13.2). Most important in this context however is that the proliferative capacity of *in vivo* aged MSC can be greatly increased when cultivating MSC at low oxygen tension [9]. These findings greatly suggest that oxygen is an important determinant of the MSC niche. In turn, occlusion and/or degeneration of the microvascular plexi within mesenchymal tissue as an accessory implication of aging may have a profound impact on the basic properties of tissue-borne MSC.

13.2.4 MSC Show Little Telomere Attrition in vivo

MSC grown in long-term culture eventually acquire a state of irreversible cell cycle arrest yet remain metabolically active and exhibit an enhanced resist-ance against apoptosis. We recognized, similarly to what has been reported by others [7, 41, 42] that the number of population doublings (PD) accumu-lated by MSC lines derived from individual donors varies considerably [10].

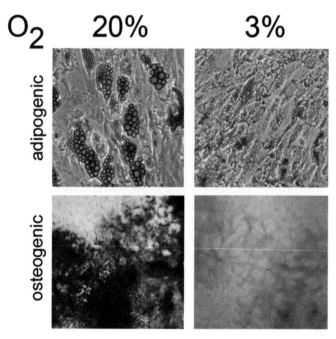

Figure 13.2 Differentiation potential of MSC towards osteogenic and adipogenic lineage. Under ambient atmospheric conditions of 20% oxygen, MSC are readily induced to differentiate into preadipocytes containing intracellular fat containing droplets which stain positive for Oil Red (upper left side) or osteoblasts that efficiently deposit minera-lizing extracellular matrix, which can be stained with Alizarin Red (lower left side). Reduced oxygen availability greatly impairs commitment to differentiate (right side) while remaining at a naïve state.

Interestingly, a good correlation between donor age and the respective PD became evident. Superficially contemplated, this circumstance might indicate that MSC may age in vivo in a similar fashion as they do in vitro. Although it is rather unclear to date what are the distinct implications of the in vivo niche with respect to cellular aging of MSC, replicative senescence as a model system was deliberately employed to sort out molecular cues of how MSC potentially age in vivo [8, 19, 43]. In fact, there are good reasons to believe that studying MSC sencescence acquired in long-term culture might be highly misleading.

For instance, in culture MSC show telomere attrition at high passages, and this type of genotoxic stress eventually contributes to the limited replicative life-span. This phenomenon can indeed be overcome by forced expression of telomerase [44]. Yet many laboratories could show that the length of telomeric ends, although being significantly higher in children [45, 46] is maintained at a considerable long length in adult age [8, 10, 47]. This suggests that expression of telomerase takes place in MSC *in vivo* be it at a very low constitutive level, or in a transient fashion resulting in maintaining the proper structure of chromosome ends. Hence changes occurring in MSC *in vivo* are only insufficiently described by patchy examinations of homogenous cultures of replicatively aging cells.

13.2.5 'Inflamm-aging' of MSC

Early in life, most, if not all, uncommitted MSC exhibit multipotential differentiation capacity, and their developmental fate is tightly controlled. Later in life however, additional types of mesenchymal precursors emerge as stable subpopulations [48–51]. Besides the rate of cell death, which is gradually increasing, MSC accumulate various forms of damage, which cannot be compensated by cellular repair [13]. We found VCAM-1 levels were found considerably increased in MSC of elderly donors [10]. MSC are greatly responsive to pro-inflammatory cytokines such as TNFα or IFNγ, which upregulate VCAM-1 transcription. As a consequence, these cells become committed to either differentiate towards the adipogenic or osteogenic lineage. In order to add-on to these results, we selected the population of VCAM1 negative MSC from elderly donors. Proliferation and self-renewal assays as well as quantitative assessment of the levels of various transcription factors involved in the regulation and control of stemness and differentiation indicated that regardless of CD106 expression, MSC exhibited apparent differences. Conclusively, a rise in VCAM1/CD106 expression is indicative for alterations of MSC stemness. Taken together, the results stronggreatly argue that due to an age-dependent chronic, systemic inflammation, a subpopulation of uncommitted MSC becomes manifest, which are predetermined mesenchymal precursors or MPC. In total, only little decline of mesenchymal progenitor cell numbers is envisaged at advanced age, the number of truly naïve, multipotential MSC however steadily declines.

13.3 Unanswered Questions Concerning the Consequences of MSC Aging

With increasing age not only is fracture incidence greatly increased, also the capacity of bone to regenerate, slows down [53]. Osteogenic differentiation is highly associated with blood vessel formation, which actually is greatly

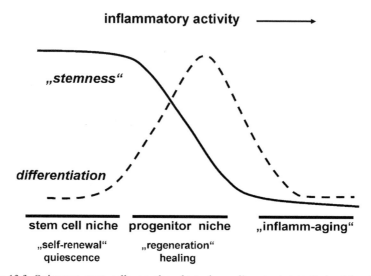

Figure 13.3 Quiescent stem cells are thought to be well protected in their niche, thereby maintaining high self-renewal potential. Heal-ing often is accompanied and could be further supported by inflammatory stimuli. Considering that daughter cells which leave the primary niche and taxi into a pro-genitor niche may become compromised by a proin-flammatory environment if these harsh conditions persist over long time or cytokine concentrations are above a certain threshold. High levels of inflammation inevitably lead to dysregulated differ-entiation of stem cell progeny; in case of chronic inflammatory deviations either systemically or topically prevailing, regenerative mechanisms become greatly confused. Such a situation has been termed 'inflamm-aging' [52].

retarded in the elderly and thus thought to be the prime cause for delayed fracture healing. Yet it remains to be determined whether debilitated MSC play a decisive role therein.

Normal wound repair is often decelerated in elderly persons or profoundly impacted in several age-related diseases which display states of non-healing or poorly healing of lesions. As MSC can be isolated from virtually all tis-sue types, it appears likely that they are involved in normal wound repair. It is however not known whether MSC simply provide daughter cells that differentiate and then directly participate in the structural repair of a wound, or whether MSC juxtaposed or residing within the wound supply paracrine factors that support wound repair and/or modulate the activity of the immune system [54]. In lieu of basic knowledge regarding their actual involvement in this process, it is furthermore highly speculative to date whether aged

MSC are confounding denominators of age-associated impairment in wound healing.

Evidence that MSC may be dysfunctional in age-associated disease is controversial. At least two reports suggest that abnormal osteogenic potential contributes to osteoporosis and that MSC growth, proliferative response, and osteogenic differentiation may be affected in this disease [55, 56]. Regarding the immunopathology of rheumatoid arthritis, it could be shown in the TNFα transgenic mouse model that MSC can migrate to the joint space, before the onset of acute inflammation, while anti-TNFα therapy actually inhibits local migration of MSC and prevents the clinical signs of arthritis [57]. Whether it is aged MSC which are all-dominantly involved in disease manifestation remains to be determined.

There is good evidence that analogous to hematopoietic stem/progenitor cells which are originators of leukemia stem cells [58], MSC and mesenchymal progenitors are trailheads of sarcomas, as sarcoma formation could potentially originate from aberrantly differentiating MSC [59]. However the fact that the prevalence of sarcomas remains low late in life greatly suggests that alterations of MSC corresponding with advancing age are not excessive enough to promote tumorigenesis. Despite these observations the concept of cancer originating from cancer stem cells (CSC) is also of interest in the context of MSC biogerontology. Needless to say that there are no experimental results available to date, which proveof whether CSC exhibit quasi multipotential differentiation potential. Also no in vivo lineage tracing could be established, which definitely clarifies a direct interrelationship of tissue-borne MSC and CSC.

Although proofs for tumor initiation by aged MSC is vague, there isare by far better evidences that MSC are involved in the progression of carcinogenesis, as MSC appear to migrate towards primary tumors and metastatic sites [60]. It is highly likely that determining chemokines, such as CCL5 or SDF-1 alpha, which are being secreted by attracted MSC actually promote emergence of metastases as shown in the case of breast cancer [61], or facilitate the recruitment of breast cancer cells to the bone marrow which was shown to be mediated through the action of CXCR4 [62]. Provided these cases of evidences, it would be tempting to examine whether chemokine expression is indeed changed in aging MSC.

13.4 Conclusions

Over recent years, information regarding MSC biology has been exploded, yet information on age-associated changes is still scarce. Due to the rising socioeconomic problems resulting from the fact that Western societies are increasingly aging, questions concerning scientific aspects of stem cell biology have becoame an emerging field of interest in the proliferating discipline of Aging Research. Understanding of how stem cells undergo age-related alterations may certainly lead to more detailed understanding of how body functions do actually change as we age. As the number of MSC therapies steadily rises, it is anticipated that this specific theme will also become increasingly important. Although pertinent questions about aging of MSC are still unanswered, some first findings refer to the corresponding niche being a confounding determining factor for well-being of this ubiquitously present stem cell type.

References

[1] M. Dominici, K. Le Blanc, I. Mueller, I. Slaper-Cortenbach, F. Marini, D. Krause, R. Deans, A. Keating, D. Prockop, and E. Horwitz. Minimal criteria for defining multipotent mesenchymal stromal cells. The International Society for Cellular Therapy position statement. *Cytotherapy*, 8(4):315–317, 2006.

[2] S. Mendez-Ferrer, T.V. Michurina, F. Ferraro, A.R. Mazloom, B.D. Macarthur, S.A. Lira, D.T. Scadden, A. Ma'ayan, G.N. Enikolopov, and P.S. Frenette. Mesenchymal and haematopoietic stem cells form a unique bone marrow niche. *Nature*, 466(7308):829–834, 2010.

[3] E.M. Horwitz and A. Keating. Nonhematopoietic mesenchymal stem cells: What are they? *Cytotherapy*, 2(5):387–388, 2000.

[4] K. Le Blanc, F. Frassoni, L. Ball, F. Locatelli, H. Roelofs, I. Lewis, E. Lanino, B. Sundberg, M.E. Bernardo, M. Remberger, G. Dini, R.M. Egeler, A. Bacigalupo, W. Fibbe, and O. Ringden. Mesenchymal stem cells for treatment of steroid-resistant, severe, acute graft-versus-host disease: a phase II study. *Lancet*, 371(9624):1579–1586, 2008.

[5] K. Le Blanc, I. Rasmusson, B. Sundberg, C. Gotherstrom, M. Hassan, M. Uzunel, and O. Ringden. Treatment of severe acute graft-versus-host disease with third party haploidentical mesenchymal stem cells. *Lancet*, 363(9419):1439–1441, 2004.

[6] C. Fehrer and G. Lepperdinger, Mesenchymal stem cell aging. *Exp. Gerontol.*, 40(12):926–930, 2005.

[7] A. Stolzing, E. Jones, D. McGonagle, and A. Scutt, Age-related changes in human bone marrow-derived mesenchymal stem cells: Consequences for cell therapies. *Mechanisms of Ageing and Development*, 129(3):163–173, 2008.

[8] W. Wagner, S. Bork, P. Horn, D. Krunic, T. Walenda, A. Diehlmann, V. Benes, J. Blake, F.-X. Huber, V. Eckstein, P. Boukamp, and A.D. Ho, Aging and replicative senescence have related effects on human stem and progenitor cells. *PloS one*, 4(6):e5846, 2009.

[9] C. Fehrer, R. Brunauer, G. Laschober, H. Unterluggauer, S. Reitinger, F. Kloss, C. Gülly, R. Gassner, and G. Lepperdinger, Reduced oxygen tension attenuates differentiation capacity of human mesenchymla stem cells and prolongs their life span. *Aging Cell*, 6:745–757, 2007.

[10] G.T. Laschober, R. Brunauer, A. Jamnig, S. Singh, U. Hafen, C. Fehrer, F. Kloss, R. Gassner, and G. Lepperdinger, Age-specific changes of mesenchymal stem cells are paralleled by upregulation of CD106 expression as a response to an inflammatory environment. *Rejuvenation Res.*, 2011.

[11] G.T. Laschober, D. Ruli, E. Hofer, C. Muck, D. Carmona-Gutierrez, J. Ring, E. Hutter, C. Ruckenstuhl, L. Micutkova, R. Brunauer, A. Jamnig, D. Trimmel, D. Herndler-Brandstetter, S. Brunner, C. Zenzmaier, N. Sampson, M. Breitenbach, K.U. Frohlich, B. Grubeck-Loebenstein, P. Berger, M. Wieser, R. Grillari-Voglauer, G.G. Thallinger, J. Grillari, Z. Trajanoski, F. Madeo, G. Lepperdinger, and P. Jansen-Durr, Identification of evolutionarily conserved genetic regulators of cellular aging. *Aging Cell*, 9(6):1084–1097, 2011.

[12] G. Laschober, R. Brunauer, A. Jamnig, S. Singh, U. Hafen, C. Fehrer, F. Kloss, R. Gassner, and G. Lepperdinger, Age-specific changes regarding osteogenesis of mesenchymal stem cells are paralleled by upregulation of VCAM1/CD106 expression as a response to an inflammatory environment. *Rejuv. Res.*, 2010. in press.

[13] G.T. Laschober, R. Brunauer, A. Jamnig, C. Fehrer, B. Greiderer, and G. Lepperdinger, Leptin receptor/CD295 is upregulated on primary human mesenchymal stem cells of advancing biological age and distinctly marks the subpopulation of dying cells. *Exp. Gerontol.*, 44(1/2):57–62, 2009.

[14] A. Stolzing and A. Scutt, Age-related impairment of mesenchymal progenitor cell function. *Aging Cell*, 5(3):213–224, 2006.

[15] C. Fehrer, C., R. Brunauer, G. Laschober, H. Unterluggauer, S. Reitinger, F. Kloss, C. Gülly, R. Gassner, and G. Lepperdinger, Reduced oxygen tension attenuates differentiation capacity of human mesenchymla stem cells and prolongs their life span. Submitted for publication, 2007.

[16] S. Sethe, A. Scutt, and A. Stolzing, Aging of mesenchymal stem cells. *Ageing Res. Rev.*, 5(1):91–116, 2006.

[17] M. Collado, M.A. Blasco, and M. Serrano, Cellular senescence in cancer and aging. *Cell*, 130(2):223–233, 2007.

[18] J.H. Chen, C.N. Hales, and S.E. Ozanne, DNA damage, cellular senescence and organismal ageing: Causal or correlative? *Nucleic Acids Res.*, 35(22):7417–7428, 2007.

[19] M.A. Baxter, R.F. Wynn, S.N. Jowitt, J.E. Wraith, L.J. Fairbairn, and I. Bellantuono, Study of telomere length reveals rapid aging of human marrow stromal cells following in vitro expansion. *Stem Cells (Dayton, Ohio)*, 22(5):675–682, 2004.

[20] A.I. Caplan, Why are MSCs therapeutic? New data: new insight. *The Journal of Pathology*, 217(2):318–324, 2009.

[21] M. Fan, W. Chen, W. Liu, G.-Q. Du, S.-L. Jiang, W.-C. Tian, L. Sun, R.-K. Li, and H. Tian, The effect of age on the efficacy of human mesenchymal stem cell transplantation after a myocardial infarction. *Rejuvenation Research*, 2010.

[22] S.V. Tokalov, S. Grüner, S. Schindler, A.S. Iagunov, M. Baumann, and N.D. Abolmaali, Molecules and a number of bone marrow mesenchymal stem cells but neither phenotype

nor differentiation capacities changes with age of rats. *Molecules and Cells*, 24(2):255–260, 2007.

[23] S. Nishida, N. Endo, H. Yamagiwa, T. Tanizawa, and H.E. Takahashi, Number of osteoprogenitor cells in human bone marrow markedly decreases after skeletal maturation. *J. Bone Miner. Metab.*, 17(3):171–177, 1999.

[24] G.F. Muschler, H. Nitto, C.A. Boehm, and K.A. Easley, Age- and gender-related changes in the cellularity of human bone marrow and the prevalence of osteoblastic progenitors. *J. Orthop. Res.*, 19(1):117–125, 2001.

[25] A.K. Majors, C.A. Boehm, H. Nitto, R.J. Midura, and G.F. Muschler, Characterization of human bone marrow stromal cells with respect to osteoblastic differentiation. *J. Orthop. Res.*, 15(4):546–557, 2001.

[26] A.I. Caplan, Adult mesenchymal stem cells for tissue engineering versus regenerative medicine. *J. Cell Physiol.*, 213(2):341–347, 2007.

[27] A. Stolzing, E. Jones, D. McGonagle, and A. Scutt, Age-related changes in human bone marrow-derived mesenchymal stem cells: Consequences for cell therapies. *Mech. Ageing Dev.*, 129(3):163–173, 2008.

[28] R.O. Oreffo, A. Bennett, A.J. Carr, and J.T. Triffitt, Patients with primary osteoarthritis show no change with ageing in the number of osteogenic precursors. *Scand. J. Rheumatol.*, 27(6):415–424, 1998.

[29] K. Stenderup, J. Justesen, E.F. Eriksen, S.I. Rattan, and M. Kassem, Number and proliferative capacity of osteogenic stem cells are maintained during aging and in patients with osteoporosis. *J. Bone Miner. Res.*, 16(6):1120–1129, 2001.

[30] J. Justesen, K. Stenderup, E.F. Eriksen, and M. Kassem, Maintenance of osteoblastic and adipocytic differentiation potential with age and osteoporosis in human marrow stromal cell cultures. *Calcif. Tissue Int.*, 71(1):36–44, 2002.

[31] G. D'Ippolito, P.C. Schiller, C. Ricordi, B.A. Roos, and G.A. Howard, Age-related osteogenic potential of mesenchymal stromal stem cells from human vertebral bone marrow. *J. Bone Miner. Res.*, 14(7):1115–1122, 2002.

[32] J. Glowacki, Influence of age on human marrow. *Calcif. Tissue Int.*, 56, Suppl 1:S50–51, 2002.

[33] G.T. Laschober, R. Brunauer, A. Jamnig, S. Singh, U. Hafen, C. Fehrer, F. Kloss, R. Gassner, and G. Lepperdinger, Age-specific changes of mesenchymal stem cells are paralleled by upregulation of CD106 expression as a response to an inflammatory environment. *Rejuvenation Res.*

[34] S.A. Kuznetsov, P.H. Krebsbach, K. Satomura, J. Kerr, M. Riminucci, D. Benayahu, and P.G. Robey, Single-colony derived strains of human marrow stromal fibroblasts form bone after transplantation in vivo. *J. Bone Miner. Res.*, 12(9):1335–1347, 1997.

[35] A. Muraglia, R. Cancedda, and R. Quarto, Clonal mesenchymal progenitors from human bone marrow differentiate in vitro according to a hierarchical model. *J. Cell Sci.*, 113 (Pt 7):1161–1166, 2000.

[36] J.D. Kretlow, Y.-Q. Jin, W. Liu, W.J. Zhang, T.-H. Hong, G. Zhou, L.S. Baggett, A.G. Mikos, and Y. Cao, Donor age and cell passage affects differentiation potential of murine bone marrow-derived stem cells. *BMC Cell Biology*, 9:60, 2008.

[37] K. Stenderup, J. Justesen, C. Clausen, and M. Kassem, Aging is associated with decreased maximal life span and accelerated senescence of bone marrow stromal cells. *Bone*, 33(6):919–926, 2003.

[38] P. Bosch, D.S. Musgrave, J.Y. Lee, J. Cummins, T. Shuler, T.C. Ghivizzani, T. Evans, T.D. Robbins, and Huard, Osteoprogenitor cells within skeletal muscle. *J. Orthop. Res.*, 18(6):933–944, 2000.

[39] G.D. Collett and A.E. Canfield, Angiogenesis and pericytes in the initiation of ectopic calcification. *Circ. Res.*, 96(9):930–938, 2000.

[40] G.Z. Eghbali-Fatourechi, J. Lamsam, D. Fraser, D. Nagel, B.L. Riggs, and S. Khosla, Circulating osteoblast-lineage cells in humans. *N. Engl. J. Med.*, 352(19):1959–1966, 2005.

[41] K.R. Shibata, T. Aoyama, Y. Shima, K. Fukiage, S. Otsuka, M. Furu, Y. Kohno, K. Ito, S. Fujibayashi, M. Neo, T. Nakayama, T. Nakamura, and J. Toguchida, Expression of the p16INK4A gene is associated closely with senescence of human mesenchymal stem cells and is potentially silenced by DNA methylation during in vitro expansion. *Stem Cells*, 25(9):2371–2382, 2007.

[42] W. Wagner, P. Horn, S. Bork, and A.D. Ho, Aging of hematopoietic stem cells is regulated by the stem cell niche. *Exp. Gerontol.*, 43(11):974–980, 2005.

[43] S. Bork, S. Pfister, H. Witt, P. Horn, B. Korn, A.D. Ho, and W. Wagner, DNA methylation pattern changes upon long-term culture and aging of human mesenchymal stromal cells. *Aging Cell*, 9(1):54–63, 2010.

[44] J.L. Simonsen, C. Rosada, N. Serakinci, J. Justesen, K. Stenderup, S.I. Rattan, T.G. Jensen, and M. Kassem, Telomerase expression extends the proliferative life-span and maintains the osteogenic potential of human bone marrow stromal cells. *Nat. Biotechnol.*, 20(6):592–596, 2002.

[45] D.M. Choumerianou, G. Martimianaki, E. Stiakaki, L. Kalmanti, M. Kalmanti, and H. Dimitriou, Comparative study of stemness characteristics of mesenchymal cells from bone marrow of children and adults. *Cytotherapy*, 12(7):881–887, 2010.

[46] K. Mareschi, I. Ferrero, D. Rustichelli, S. Aschero, L. Gammaitoni, M. Aglietta, E. Madon, and F. Fagioli, Expansion of mesenchymal stem cells isolated from pediatric and adult donor bone marrow. *J. Cell Biochem.*, 97(4):744–754, 2006.

[47] T.C. Lund, A. Kobs, B.R. Blazar, and J. Tolar, Mesenchymal stromal cells from donors varying widely in age are of equal cellular fitness after in vitro expansion under hypoxic conditions. *Cytotherapy*, 12(8):971–981, 2011.

[48] G. Lepperdinger, Open-ended question: Is immortality exclusively inherent to the germline? – A mini-review. *Gerontology*, 55(1):114–117, 2009.

[49] G. Lepperdinger, Aging stem cells and regenerative biomedicine: concepts, opportunities and technological advances. *Exp. Gerontol.*, 43(11):967, 2008.

[50] G. Lepperdinger, R. Brunauer, A. Jamnig, G. Laschober, and M. Kassem, Controversial issue: Is it safe to employ mesenchymal stem cells in cell-based therapies? *Exp. Gerontol.*, 43(11):1018–1023, 2008.

[51] G. Lepperdinger, P. Berger, M. Breitenbach, K.U. Frohlich, J. Grillari, B. Grubeck-Loebenstein, F. Madeo, N. Minois, W. Zwerschke, and P. Jansen-Durr, The use of genetically engineered model systems for research on human aging. *Front Biosci.*, 13:7022–7031, 2008.

[52] C. Franceschi, M. Capri, D. Monti, S. Giunta, F. Olivieri, F. Sevini, M.P. Panourgia, L. Invidia, L. Celani, M. Scurti, E. Cevenini, G.C. Castellani, and S. Salvioli, Inflammaging and anti-inflammaging: A systemic perspective on aging and longevity emerged from studies in humans. *Mech. Ageing. Dev.*, 128(1):92–105, 2007.

[53] R. Gruber, H. Koch, B.A. Doll, F. Tegtmeier, T.A. Einhorn, and J.O. Hollinger, Fracture healing in the elderly patient. *Exp. Gerontol.*, 41(11):1080–1093, 2006.

[54] Y. Wu, L. Chen, P.G. Scott, and E.E. Tredget, Mesenchymal stem cells enhance wound healing through differentiation and angiogenesis. *Stem Cells*, 25(10):2648–2659, 2007.

[55] L. Dalle Carbonare, M.T. Valenti, M. Zanatta, L. Donatelli, and V. Lo Cascio, Circulating mesenchymal stem cells with abnormal osteogenic differentiation in patients with osteoporosis. *Arthritis Rheum.*, 60(11):3356–3365, 2009.

[56] J.P. Rodriguez, S. Garat, H. Gajardo, A.M. Pino, and G. Seitz, Abnormal osteogenesis in osteoporotic patients is reflected by altered mesenchymal stem cells dynamics. *J. Cell Biochem.*, 75(3):414–423, 1999.

[57] L. Marinova-Mutafchieva, R.O. Williams, K. Funa, R.N. Maini, and N.J. Zvaifler, Inflammation is preceded by tumor necrosis factor-dependent infiltration of mesenchymal cells in experimental arthritis. *Arthritis Rheum.*, 46(2):507–513, 2002.

[58] E. Passegué, C.H.M. Jamieson, L.E. Ailles, and I.L. Weissman, Normal and leukemic hematopoiesis: Are leukemias a stem cell disorder or a reacquisition of stem cell characteristics? *PNAS*, 100 Suppl:11842–11849, 2003.

[59] N. Tang, W.-X. Song, J. Luo, R.C. Haydon, and T.-C. He, Osteosarcoma development and stem cell differentiation. *Clinical Orthopaedics and Related Research*, 466(9):2114–2130, 2008.

[60] G. Lazennec, and C. Jorgensen, Concise review: adult multipotent stromal cells and cancer: Risk or benefit? *Stem Cells*, 26(6):1387–1394, 2008.

[61] A.E. Karnoub, A.E., A.B. Dash, A.P. Vo, A. Sullivan, M.W. Brooks, G.W. Bell, A.L. Richardson, K. Polyak, R. Tubo, and R.A. Weinberg, Mesenchymal stem cells within tumour stroma promote breast cancer metastasis. *Nature*, 449(7162):557–563, 2007.

[62] K.E. Corcoran, K.A. Trzaska, H. Fernandes, M. Bryan, M. Taborga, V. Srinivas, K. Packman, P.S. Patel, and P. Rameshwar, Mesenchymal stem cells in early entry of breast cancer into bone marrow. *PloS one*, 3(6):e2563, 2008.

14

Cardiac Regeneration: Current View and Future Outlook

Konstantina-Ioanna Sereti, Angelos Oikonomopoulos
and Ronglih Liao

Cardiac Muscle Research Laboratory, Cardiovascular Division, Department of Medicine, Brigham and Women's Hospital and Harvard Medical School, Boston, MA, USA; e-mail: rliao@rics.bwh.harvard.edu

Abstract

Cardiovascular diseases represent the leading cause of mortality worldwide. Even though there have been continuous advances in treatment options for cardiovascular patients, new therapeutic strategies focusing on replenishing the lost myocardium are required.

A large number of studies, using various methodologies, have led to a widely accepted notion supporting that the adult mammalian heart possesses regenerative potential. In line with this notion is the identification of resident cardiac stem/progenitor cells able to regenerate cardiac tissue. These cells represent a promising tool for the development of cell based therapies. Recent clinical trials using resident cardiac progenitors have generated very promising initial results.

Ongoing work combining studies in both animals and humans will allow us to fully exploit the therapeutic potential of cardiac stem/progenitor cells and cardiac regeneration

Keywords: myocardial infarction (MI), cardiomyocytes, stem/progenitor cells, regeneration.

P. Di Nardo (Ed.), Adult Stem Cell Standardization, 183–190.

14.1 Introduction

Cardiovascular disease represents the leading cause of death worldwide [15], and in the United States alone, the 2010 total hospital costs for cardiovascular events was projected to reach $155,700,000 [14]. Despite ongoing advances in the care of cardiovascular patients, for those patients suffering from end stage heart failure, to date, cardiac transplantation and mechanical ventricular assist devices remain the sole options for prolonging life. These treatments are complicated by the degree of invasiveness, as well as limited availability of donor organs, immune rejection issues and mechanical complications; as such, new therapeutic strategies are required.

A major cause of heart failure is myocardial infarction (MI). During an MI, a considerable number of cardiomyocytes is lost [8] and the heart is unable to compensate adequately for this loss, leading to further damage of the surviving myocardium and subsequent overall tissue dysfunction. The concept of replenishing lost cardiac cells with cells capable of developing into *de novo* functional cardiac cells, while challenging, is certainly enticing. While the field of cardiac regeneration is still relatively young, several key observations have been made over the past decade, raising the promise of such future therapeutic options. These observations include the ability of bone marrow derived stem cells to undergo trans-differentiation into various cells in the heart including cardiomyocytes and partially repair/regenerate infarcted myocardium [7], as well as several lines of evidence now suggesting the existence of resident cardiac stem/progenitor cells in adult heart tissue. Moreover, compelling evidence suggests that new cardiomyocytes may be generated in adult cardiac tissue. Given the large array of reviews and commentaries which have been published on this topic, this chapter will focus on the current status as well as future perspectives in the field of cardiac regeneration.

14.2 The Adult Heart is Endowed with Regenerative Capacity

For decades, the adult heart has been considered a terminally differentiated organ with no ability for self-renewal. As such, it has been widely believed that the number of cardiomyocytes is established shortly after birth and any cells lost to either the normal aging process or as a consequence of cardiac injury cannot be replaced. This concept, however, has recently been challenged and a large body of current work has been conducted to address this very issue. In 2001, Beltrami et al. proposed that the human heart possesses regenerative potential based on the observation of dividing cardiomyocytes in human hearts obtained from patients after myocardial infarction [2]. The authors reported that 4% of the cardiomyocytes found in the infarct border zone expressed Ki67, a marker for cycling cells, and this percentage was reduced to 1% in the remote region. Ki67 labeling was accompanied by cytokinesis, further demonstrating cardiomyocyte proliferation. Interestingly, cycling cardiomyocytes were also detected in healthy hearts albeit at a far lower frequency (0.11%). Similarly, using genetic fate mapping and transgenic mice expressing GFP only in cardiomyocytes, following tamoxifen treatment, Hsieh and colleagues were able to confirm that cardiomyocyte regeneration does occur with cardiac stressors, including myocardial infarction or aortic constriction [10]. The authors found that the percentage of GFP positive cardiomyocytes decreased by 15.3% in the infarct border zone and by 7.1% in the remote area suggesting *de novo* cardiomyocyte generation following injury. During normal development, however, the percentage of GFP positive cardiomyocytes remained unchanged, suggesting that no significant *de novo* cardiomyocyte formation had occurred.

More recently, additional evidence for cardiac regeneration has also been provided by the work of Bergmann and colleagues using a C14 dating method [4]. Briefly, the basis of C14 dating compares the relative age of cardiomyocytes to an individual's biological age. The abundance of above ground nuclear testing in the 1950s and the resulting transient increase in atmospheric C14, allows for such relative dating. With mathematical modeling and calculation, it was estimated that approximately 1% of cardiomyocytes were "younger" than biological age in individuals up to 22 years old and 0.5% in people up to an age of 55. Albeit at a rather low percentage, these results demonstrated that cardiomyocytes are renewed in an adult heart, and cumulatively during an average life span, approximately 50% of cardiomyocytes will be replaced by new cells [4]. Interestingly, a more recent study by Kajstura et al. has taken advantage of post-mortem samples from cancer patients who received radiosensitizer iododeoxyridine (IdU) treatments [11]. This nucleotide analogue incorporates into the DNA while the cell is undergoing mitosis thus allowing the detection of newly formed cells. The authors conclude that the adult human cardiac cells (myocytes and non-myocytes) are replaced several times during a lifetime with an average of 22% of cardiomyocytes being generated per year. This estimate, while still much higher than other reported data, is consistent with previous work published by this group using different methods [2].

Based on the past decade of scientific literature, a general consensus has been reached supporting the notion that the adult mammalian heart does, in fact, have regenerative potential, particularly following cardiac injury. This observation has been demonstrated in rodents, large animals, and humans using various methodologies, ranging from genetic fate mapping to immunohistochemistry. While the absolute degree of regeneration is still contested and remains an open question, the fact that this regeneration occurs at all is widely supported. Additionally, it is unclear whether newly formed cardiomyocytes originate from existing cells which re-enter the cell cycle and divide, are *de novo* generated from resident or extra-cardiac stem/progenitor cells, or if it is a combination of the two; this is currently under extensive investigation. In contrast to the disease or injury state, the ability of the heart to regenerate under basal conditions as well as during normal aging remains largely controversial; current studies report contradictory findings ranging from a sizeable fraction of total cells to be newly formed cardiomyocytes to none at all, depending on the laboratory and methodologies employed [2, 4, 10, 13].

14.3 The Adult Heart Contains Resident Cardiac Stem-Progenitor Cells

Supporting the notion that the adult heart is capable of regeneration is the recent identification of resident cardiac stem/progenitor cells in adult myocardium. Several laboratories have confirmed the existence of these cardiac stem/progenitor cells through the expression of cell surface markers, including c-kit [3] and Sca1 [18], as well as surface marker-independent isolation methods such as cardiosphere formation [16] and specific cellular efflux of the DNA binding dye Hoechst (side population cells) [9]. The common feature of all of these stem/progenitor cell populations is that they can differentiate into functional cardiomyocytes, smooth muscle cells and endothelial cells upon proper stimulation. It still remains unclear however, whether these populations represent separate pools of stem cells or a common cell at varying stages of differentiation. The discovery of resident cardiac stem/progenitor cells has brought great excitement to the field and has boosted the hope of cell based therapies for cardiac failure.

Resident cardiac stem/progenitor cells represent a very promising opportunity for the development of cell based therapies for cardiac disease. Isolated cardiac stem/progenitor cells have been shown to survive and exert beneficial effects when implanted into injured hearts, *in vivo* [3, 6, 19]. To date, however, the vast majority of these studies have been performed in small animal models [1]. cKit$^+$ cardiac stem cells and cardiosphere progenitor cells are to date the only two cell types which have been isolated from human patient biopsies, expanded in culture, and subsequently transplanted back to the same patient's injured myocardium. Phase I safety and feasibility clinical trials are currently ongoing for both cKit$^+$ cardiac stem cells and cardiosphere progenitor cells. In particular, these studies will evaluate the feasibility of isolating and expanding functional cardiac stem/progenitor cells from humans with underlying cardiovascular disease, prior to myocardial injury and cardiac dysfunction. Recently, very promising initial results from the cKit$^+$ cardiac stem cell trial were reported by Bolli et al. [5] in patients with LV dysfunction due to myocardial infarction. cKit$^+$ cells harvested during CABG (Coronary Artery Bypass Graft) surgery were re-introduced intracoronarily four months after surgery. Results from six patients, four months after treatment, revealed significant improvement in LV systolic function and overall functional capacity as well as reduced infarct size. Although preliminary, these results are very encouraging and the outcome of this, and other similar trials, is anticipated with great interest.

14.4 Future Outlook

Therapeutic cardiac regeneration may be achieved through two approaches: supplementing injured hearts with implanted stem/progenitor cells or activating the regenerative capacity of the endogenous cardiac stem/progenitor cell pool. For the past decade, various exogenous extra-cardiac stem/progenitor cells have been extensively investigated not only in animal models but recently in humans in several clinical trials. To date, these human studies have reported a marginal benefit of cell therapy on top of current medical treatment, though several important lessons have been learned and will provide valuable insight into clinical studies. There are several major questions and areas of uncertainty in the field of cardiac regeneration that require our continued attention. The most advantageous cell type still remains an open question. Towards this, recent trials using adult cardiac cKit cells and cardiosphere-derived progenitor cells may provide potentially exciting information. Given the harsh inflamed local environment present in recently injured myocardium, issues such as the optimal timing for delivery of cells as well as which patients will benefit most from cell therapy remain unclear. Manipulation of cells to maximize survival and differentiation following engraftment also remains an important issue. The above mentioned as well as other areas are ongoing subjects of investigation worldwide. Only with both basic science studies in animals and carefully planned human studies will these questions be addressed and the true potential of cell based therapy and cardiac regeneration be realized.

The existence of resident cardiac stem/progenitor cells in adult heart is currently a concept broadly accepted, however these resident stem/progenitor cells seem unable to adequately regenerate the lost myocardium after injury. It has been reported that cardiac stem cells possess proliferation capacity which is increased following injury. Mouquet et al. demonstrated that CSP cells decrease one day after MI but rapidly reach initial levels within 7 days [17]. This increase in cell numbers is accompanied by increased levels of the proliferation marker Ki67, but also with decreased capacity for cardiomyogenic differentiation. Moreover, cKit$^+$ cells isolated from patients with heart failure were 4-fold increased compared to healthy hearts [12]. These findings demonstrate that the equilibrium between proliferation and differentiation of cardiac stem cells after injury may be deregulated. There is a clear proliferative response to the stress signals; however the corresponding generation of new cardiomyocytes may be hindered. Further investigation of the molecular mechanisms that control such events are imperative in order to manipulate these biological processes for therapeutic regeneration. It may be even possible to stimulate endogenous stem/progenitor cells for such regeneration. As an additional hurdle, injured myocardium represents an unfriendly environment to both endogenous and exogenously delivered stem/progenitor cells. *In vivo* or *ex vivo* treatment of cells with various molecules and manipulation of pathways that control cell survival, adherence, migration as well as the switch between proliferation and differentiation holds great potential for the eventual selection of the optimal cell population for cardiac regeneration.

14.5 Closing Remarks

Cardiac regeneration is an attractive new therapeutic option that has garnered increasing enthusiasm. The identification of resident cardiac progenitor cells able to proliferate and differentiate upon proper stimulation represents a powerful tool for cell based therapies. Results from *in vitro* as well as *in vivo* animal studies have further revealed the immense potential of these cells. Further elucidation of the mechanisms that control stem cell survival, engraftment and differentiation into myocardial cells is imperative in order to utilize these populations to their full potential. Clinical trials using resident cardiac stem/ progenitor cells are expected to provide critical information that will help us decipher the therapeutic potential of cardiac regeneration. Moreover, it is of great importance to develop methods to either manipulate stem cells (genetic manipulation, pre-conditioning) or reprogram somatic cells (iPS) that will allow us to generate powerful stem cell populations with controlled characteristics. While the use of cardiac stem/progenitor cells for therapeutic regeneration is largely conceptual in its current stage, ongoing work in this area has rapidly propelled this concept closer to reality.

References

[1] Barile, L., Messina. E., Giacomello, A. and Marban, E. Endogenous cardiac stem cells. *Prog. Cardiovasc. Dis.*, 50:31–48, 2007.

[2] Beltrami, A.P., Urbanek, K., Kajstura, J., Yan, S.M., Finato, N., Bussani, R., Nadal-Ginard, B., Silvestri, F., Leri, A., Beltrami, C.A. and Anversa, P. Evidence that human cardiac myocytes divide after myocardial infarction. *N. Engl. J. Med.*, 344:1750–1757, 2001.

[3] Beltrami, A.P., Barlucchi, L., Torella, D., Baker, M., Limana, F., Chimenti, S., Kasahara, H., Rota, M., Musso, E., Urbanek, K., Leri, A., Kajstura, J., Nadal-Ginard, B. and Anversa, P. Adult cardiac stem cells are multipotent and support myocardial regeneration. *Cell*, 114:763–776, 2003.

[4] Bergmann, O., Bhardwaj, R.D., Bernard, S., Zdunek, S., Barnabe-Heider, F., Walsh, S., Zupicich, J., Alkass, K., Buchholz, B.A., Druid, H., Jovinge, S. and Frisen, J. Evidence for cardiomyocyte renewal in humans. *Science*, 324:98–102, 2009.

[5] Bolli, R., Chugh, A.R., D'Amario, D., Stoddard, M.F., Ikram, S., Wagner, S.G., Beache, G.M., Leri, A., Hosoda, T., Loughran, J.H., Goihberg, P., Fiorini, C., Solankhi, N.K., Fahsah, I., Chatterjee, A., Elmore, J.B., Rokosh, D.G., Slaughter, M.S., Kajstura, J. and Anversa, P. Use of Cardiac Stem Cells for the Treatment of Heart Failure: Translation from Bench to the Clinical Setting. Circ Res Late-Breaking Basic Science Abstracts from the American Heart Association's Scientific Sessions 2010, Chicago, Illinois, Abstract 23734, 2010.

[6] Dawn, B., Stein, A.B., Urbanek, K., Rota, M., Whang, B., Rastaldo, R., Torella, D., Tang, X.L., Rezazadeh, A., Kajstura, J., Leri, A., Hunt, G., Varma, J., Prabhu, S.D., Anversa, P. and Bolli, R. Cardiac stem cells delivered intravascularly traverse the vessel barrier, regenerate infarcted myocardium, and improve cardiac function. *Proc. Natl. Acad. Sci. USA*, 102:3766–3771, 2005.

[7] Dimmeler, S., Burchfield, J. and Zeiher, A.M. Cell-based therapy of myocardial infarction. *Arterioscler. Thromb. Vasc. Biol.*, 28:208–216, 2008.

[8] Dorn, G.W. II. Having a change of heart: Reversing the suicidal proclivities of cardiac myocytes. *Trans. Am. Clin. Climatol. Assoc.*, 120:189–198, 2009.

[9] Hierlihy, A.M., Seale, P., Lobe, C.G., Rudnicki, M.A. and Megeney, L.A. The post-natal heart contains a myocardial stem cell population. *FEBS Lett.*, 530:239–243, 2002.

[10] Hsieh, P.C., Segers, V.F., Davis, M.E., MacGillivray, C., Gannon, J., Molkentin, J.D., Robbins, J. and Lee, R.T. Evidence from a genetic fate-mapping study that stem cells refresh adult mammalian cardiomyocytes after injury. *Nat. Med.*, 13:970–974, 2007.

[11] Kajstura, J., Urbanek, K., Perl, S., Hosoda, T., Zheng, H., Ogorek, B., Ferreira-Martins, J., Goichberg, P., Rondon, C., D'Amario, D., Rota, M., Del Monte, F., Orlic, D., Tisdale, J., Leri, A. and Anversa, P. Cardiomyogenesis in the adult human heart. *Circ. Res.*, 2010.

[12] Kubo, H., Jaleel, N., Kumarapeli, A., Berretta, R.M., Bratinov, G., Shan, X., Wang, H., Houser, S.R. and Margulies, K.B. Increased cardiac myocyte progenitors in failing human hearts. *Circulation*, 118:649–657, 2008.

[13] Linke, A., Muller, P., Nurzynska, D., Casarsa, C., Torella, D., Nascimbene, A., Castaldo, C., Cascapera, S., Bohm, M., Quaini, F., Urbanek, K., Leri, A., Hintze, T.H., Kajstura, J. and Anversa, P. Stem cells in the dog heart are self-renewing, clonogenic, and multipotent and regenerate infarcted myocardium, improving cardiac function. *Proc. Natl. Acad. Sci. USA*, 102:8966–8971, 2005.

[14] Lloyd-Jones, D., Adams, R.J., Brown, T.M., Carnethon, M., Dai, S., De Simone, G., Ferguson, T.B., Ford, E., Furie, K., Gillespie, C., Go, A., Greenlund, K., Haase, N., Hailpern, S., Ho, P.M., Howard, V., Kissela, B., Kittner, S., Lackland, D., Lisabeth, L., Marelli, A., McDermott, M.M., Meigs, J., Mozaffarian, D., Mussolino, M., Nichol, G., Roger, V.L., Rosamond, W., Sacco, R., Sorlie, P., Stafford, R., Thom, T., Wasserthiel-Smoller, S., Wong, N.D. and Wylie-Rosett, J. Executive summary: heart disease and stroke statistics – 2010 update: A report from the American Heart Association. *Circulation*, 121:948–954, 2010.

[15] Mathers, C.D., Boerma, T. and Ma Fat, D. Global and regional causes of death. *Br. Med. Bull.*, 92:7–32, 2009.

[16] Messina, E., De Angelis, L., Frati, G., Morrone, S., Chimenti, S., Fiordaliso, F., Salio, M., Battaglia, M., Latronico, M.V., Coletta, M., Vivarelli, E., Frati, L., Cossu, G. and Giacomello, A. Isolation and expansion of adult cardiac stem cells from human and murine heart. *Circ. Res.*, 95:911–921, 2004.

[17] Mouquet, F., Pfister, O., Jain, M., Oikonomopoulos, A., Ngoy, S., Summer, R., Fine, A. and Liao, R. Restoration of cardiac progenitor cells after myocardial infarction by self-proliferation and selective homing of bone marrow-derived stem cells. *Circ. Res.*, 97:1090–1092, 2005.

[18] Oh, H., Chi, X., Bradfute, S.B., Mishina, Y., Pocius, J., Michael, L.H., Behringer, R.R., Schwartz, R.J., Entman, M.L. and Schneider, M.D. Cardiac muscle plasticity in adult and embryo by heart-derived progenitor cells. *Ann. N.Y. Acad. Sci.*, 1015:182–189, 2004.

[19] Oyama, T., Nagai, T., Wada, H., Naito, A.T., Matsuura, K., Iwanaga, K., Takahashi, T., Goto, M., Mikami, Y., Yasuda, N., Akazawa, H., Uezumi, A., Takeda, S. and Komuro, I. Cardiac side population cells have a potential to migrate and differentiate into cardiomyocytes in vitro and in vivo. *J. Cell. Biol.*, 176:329–341, 2007.

15

Adipose-Derived Stem Cells as Source for Tissue Repair and Regeneration

Maria Prat, Stefano Pietronave and Andrea Zamperone

Department of Medical Sciences, Università del Piemonte Orientale "A. Avogadro", Novara, Italy; e-mail: maria.prat@med.unipmn.it

Abstract

Stem cells are present in all adult tissues, although with great quantitative and qualitative differences, which correlate with the specific tissue repair potential and, within each tissue, they are generally localized in defined sites or "niches". Mesenchymal stem cells were first isolated from bone marrow stroma, and were shown to be able to differentiate into lineages of tissues of mesodermal origin, such as skeletal muscle, bone, tendon, cartilage, and fat, under appropriate culturing conditions. Since the last 10 years mesenchymal stem cells have been detected also in adipose tissue. They show the same properties with the added advantages that they can be harvested easily and repeatedly in higher amounts, with lower patient discomfort and risk of morbidity. Due to their proliferation and differentiation plasticity they are thus a promising tool for tissue regeneration. The isolation, characterization, preclinical and clinical application of adipose-derived stem cells (ASCs) are reviewed in this article.

Keywords: adipose tissue, adult stem cells, bone marrow stromal cell, differentiation, tissue engineering.

P. Di Nardo (Ed.), Adult Stem Cell Standardization, 191–208.

15.1 Introduction

Stem cells (SC) become more and more important for regenerative medicine and tissue engineering because of their main properties: self-renewal and multipotency [1, 2].

While Embyo Stem Cells (ESC) and induced Pluripotent Stem (iPS) cells should have the greatest therapy potential, they present limits on the ground of ethical considerations, safety, cell regulation, genetic manipulation and costs [3–5]. For these reasons, adult SCs are a preferentially and more widely used option [6]. Indeed, not only are they involved in rebuilding tissues damaged upon pathological events, but they are also responsible of body tissues homeostasis in physiological conditions. They are present in all tissues, generally in the so- called niches, although at different levels, depending on the properties and functions of the each specific tissue. They differ also for their stemness level and differentiation potential, In most cases they are committed, i.e. they will develop the phenotype of the tissue in which they are embedded. SCs present in tissues of mesodermal origin are called Mesenchymal Stem Cells (MSC). Historically they were identified in the bone marrow more than 40 years ago [7]; since the last 10 years they have been detected also in adipose tissue [8]. This tissue seems to be an even better option, since, besides fulfilling all the attributes of bone marrow derived SCs (namely self-renewal, multipotency and autologous source), it is endowed with additional advantages. Indeed SCs can be harvested repeatedly in higher amounts, with lower patient discomfort and risk of morbidity, because of a minimally invasive procedure [9–11]. Moreover, they can be easily enriched by simple plating on a plastic surface and appear to be more genetically stable in long-term culture [12]. It is calculated that from a liposuction (ranging from 100 ml to 3 lt) an average of $375 \times 10E3$ cells/ml can be recovered within 4–5 days in tissue culture [13]. For all the above reasons adipose tissue has become an alternative attractive source of SCs.

Adipose tissue consists of different cell populations, namely adipocytes, pre-adipocytes, endothelial cells and their progenitors, pericytes, and vascular smooth muscle cells. SCs are part of the so-called stromal vascular fraction (SVF). The acronym MSC stands for both mesenchymal stromal cells and mesenchymal stem cells and should be defined in each work, as suggested by the International Society and Tissue Stem Cell Therapy (ISCT) [14, 15]. Moreover, as in rapidly evolving fields, different names have been used to describe these cells derived from adipose tissue; now agreement has been reached under proposal of the International Fat Applied Technology Society

(IFATS): the term ASCs, for Adipose-derived Stem Cells, defines the isolated, plastic-adherent, multipotent cells, which express specific cell surface markers [2, 14].

15.2 Isolation and Proliferation of Adipose Stem Cells

Adipose tissue is present in different organism districts, and on this basis we can distiguish subcutaneous, thoracic and visceral fat [9]. Also the epicardium contains adipose tissue and indeed it was found to contain SCs, which due to their close proximity to myocardium appear to have a role in development, cardiac diseases and repair [16]. For the purposes of regenerative medicine ASCs are generally isolated from subcutaneous depots, which are easily accessible and, due to the increased incidence of obesity in western countries, are abundantly available. In humans ASCs can be isolated from fat tissue wastes resulting from plastic surgery, as liposuction aspirates (tumenescent or ultrasound-assisted), or from reconstructive surgery, as resected tissue fragments. In the latter case these are finely minced, before undergoing through type I collagenase digestion, centrifugation, and filtration on 100 μm nylon mesch [10]. Centrifugation allows the elimination of the floating mature adipocytes and the recovery of a pellet containing the SVF components, including ASCs besides cells of the hematopoietic lineages. After lysis of the possible contaminating erythrocytes, ASCs are enriched on the basis of their propensity to adhere to the plastic surface of the tissue culture plates, where they are seeded at a density of about 150,000/cm^2 in DMEM-10% FBS and antibiotics for expansion and characterization. The term PLA for processed lipoaspirate is used to define cells at this stage of manipulation [9, 10].

Due to the limited, and somewhat contradictory, number of reports, it is not clear yet if a peculiar harvesting technique or a harvesting site can result in better yields. Apparently the number of clonogenic potential could be affected by the harvesting site. The age of the donor can also have an effect. Indeed it has been reported that the number of SCs decreases with age and their potential is more prone to the adipogenic phenotype [17]. ASCs can be expanded and maintained in tissue culture for at least 3 months and seven passages [18, Pietronave and Zamperone, personal observation]. Although human ASC seem to be refractory to cell transformation, extreme caution should be used in their manipulation and translational application, since spontaneous transformation has been reported for murine MSCs [19, see also below]. To speed up and standardize the procedure commercial bench top closed systems, such as Cytori's Celution^{TM} [20], already approved by the US FDA [21], and Tissue Genesis TGI 1000 platform [22] have been developed.

Once cells are isolated, a detailed molecular characterization must be performed.

15.3 Characterization of Adipose Stem Cells

15.3.1 Immunophenotype of Adipose Stem Cells

Isolated, undifferentiated ASCs cannot be identified by a single surface marker or a few of them, but a panel of markers is needed. In general the same markers originally used for SCs derived from bone marrow are used. Indeed only minor differences have emerged, when SCs from the two sources have been compared [23–25] (Table 15.1).

There is general consensus on the following markers: CD105, CD90, CD73, CD29, CD44, CD13, CD166 ASCs should be negative or express at low levels markers of the hematopoietic and angiogenic lineages, such as CD45, CD34, CD31, CD11b. Moreover ASCs are negative for CD133 and MHC class II molecules. Other markers with moderate expression are CD9, CD49b, CD146 and STRO-1. The markers expressed are mainly adhesion molecules (CD9, CD29, CD49b and CD105), receptor molecules (CD44), extracellular matrix molecules (CD90, CD146). The expression of some markers, such as CD34, CD106, STRO-1, is still controversial and may be ascribed to many reasons, mainly to differences in the antibodies used and the detection methods (cyofluorimetry or immunohistochemistry and immunofluorescence), but also to donor heterogeneity.

Moreover, the fact that the immunophenotype profile changes as a function of time in culture and plastic adherence must be taken into consideration. Indeed the expression of markers such as CD29, CD44, CD73, CD90 is increased in the first passages and then is stabilized at high levels [26]. In the meantime hematopoietic cell markers, already expressed at low level, decrease significantly or are lost. This finding probably reflects the fact that cells are selected for their adherence to plastic and thus a more homogeneous cell population emerges from the original SVF. The expression of some markers can also be affected by the type of culture media, namely serum, which can be of fetal calf or human origin or serum free conditions [2]. For all the above reasons, while it is nearly impossible to unify protocols for immunophenotype characterization, it is recommended that the minimal criteria discussed above should be followed.

Table 15.1 Phenotypic characterization of human adipose stem cells.

	ASC	BM-MSC[a]	UCB[a]
CD90	++	++	++
CD73	++	++	++
CD29	++	++	++
CD44	++	++	++
CD105	++	++	++
CD9	+	nd	nd
CD49b	+	nd	nd
CD146	+	nd	nd
STRO-1	+	nd	nd
CD13	+	nd	nd
CD166	+	nd	nd
CD34	+/−	+/−	+/−
CD45	−	+/−	+/−
CD31	−	nd	nd
CD11b	−	nd	nd
CD133	−	+/−	+/−
MHC class II	−	+/−	−

Adapted from [2, 9–11, 13, 18, 25, 26, 63, 69, 71, 85].
[a]From [85] (similar to other authors).
++ = high expression; + = moderate expression, +/− = low or no expression.

15.3.2 Differentiation Potential of Adipose Stem Cells

Based on studies of developmental biology it is clear that MSC have the ability to differentiate towards many different cell types. The accepted and conventional minimal criteria to define ASCs requires that they could be induced to differentiate towards the chondrogenic, osteogenic, and adipogenic phenotypes. To reach a fully differentiated phenotype cells must generally receive a combination of stimuli of different type, trying to recapitulate the *in vivo* situation and this is a major task, quite difficult to reproduce *in vitro*. Indeed stimuli can be of a biophysical type, such as mechanical, electrical, topographical, or of a chemical nature, such as soluble growth/differentiation factors, hormones, small defined bioactive molecules as vitamins. Most of the work is carried on using the latter class of stimuli, because it is less complicated. It is clear from these premises that it is difficult to obtain a fully differentiated cell population. However, since differentiation proceeds through sequential steps and many differentiation markers can trace its evolution, their detection is a sign of the ongoing differentiation. Protocols for differentiation may differ somehow among the laboratories, but general principles are shared.

Adipogenic differentiation. This is the natural destiny of ASCs, due to the microenvironment in which they are located. ASCs can be induced to adipocytes with chemically defined serum free medium containing insulin or IGF-1, triiodothyronine and transferring, after a brief phase in serum promoting their adhesion to the plastic substrate [27, 28]. Other additives are isobutylmethylxanthine, hydrocortisone or dexamethasone, indomethacin. After one week treatment, neutral lipid containing vacuole are detectable by Oil Red O or Nile Red staining, as well as many adipogenic mRNAs (LPL, PPAR-c, C/EBP) [29].

Osteogenic differentiation. Glucocorticoids (for example dexamethasone), ascorbate, vitamin D_3, β-glycerophosphate, and a series of soluble factors belonging to the TGF-β superfamily, called bone morphogenetic proteins (BMPs) are used for this purpose. The induced commitment and differentiation can then be evaluated on the basis of the expression of a series of markers, such as type I collagen, alkaline phosphatase, osteopontin, osteocalcin, bone sialoproteins, RUNX2, BMP-2, BMP-4, parathyroid hormone-related protein, and their receptors [30]. For long term treatment also mineralization can be detected by Alizarin Red or von Kossa staining [31].

Chondrogenic differentiation. This is routinely carried on in micro mass cultures or pellet culture systems, conditions more closely mimicking the *in vivo* pre-cartilage condensation during embryonic development [32]. Ascorbate, dexamethasone, L-proline and TGF-β3, as well as BMP-6 are the required specific soluble molecules in media. Differentiation is monitored by the expression of type II and type VI collagens, and aggrecan [33].

Detailed comparison of the differentiation potential of ASCs and bone marrow derived mesenchymal stem cells towards the above three phenotypes has been performed at transcriptome level [34]. Considerable similarities have been uncovered. In both cases the same pathways seem to be involved with a bifurcation into a common one for osteogenesis and adipogenesis and a separate one for chondrogenesis. Small differences can be detected in the maturation steps leading to fully differentiated cells. In particular stem cells from bone marrow have a higher propensity to differentiate towards bone and cartilage, while ASCs differentiate better into adipocytes.

15.3.3 Differentiation Towards Other Phenotypes

In addition to trilineage differentiation capacity, as the more studied bone marrow derived mesenchymal stem cells, ASCs have the ability to differentiate also in other cell types of mesodermal origin, such as cardiomyocytes, skeletal muscle cells, smooth muscle cells, and endothelial cells.

To induce cardiomyocyte differentiation, ASCs were generally treated with the demethylating agent 2-deoxy-5-azacytidine; later Planat-Benard reported spontaneous differentiation of murine ASCs [35]. Except one study [36], the percentage of cells acquiring differentiation markers or activity such as spontaneous beating, was low [37, 38]. Relatively few experimental studies have been performed *in vivo* in different animal models, in which transplanted cells showed some amelioration of cardiac functions after the induced infarction [37, 39–41], the main problem probably being related to the low survival and engraftment of the transplanted cells, besides the complexity of the myocardial infarct models [9].

Under appropriate induction conditions ASCs can differentiate towards the skeletal myocyte phenotype, as evident on the basis of the expression of specific markers, such as myoD, myogenin, myosin heavy chain, as well as on morphological criteria, such as cell fusion and formation of multi-nucleated myotubes [42, 43].

ASCs have also a considerable proangiogenic potential, regarding vessel incorporation [44–46], post-ischemic neovascularization [47], although these activities may probably be ascribed more to the secretion of proangiogenic factors [48] than to the direct differentiation to endothelial cells. ASCs have a profound potential to stimulate morphogenesis of endothelial cells into branching networks of cord structures [49], can be induced to smooth muscle cell differentiation by a series of factors, and their use in blood vessel engineering is under active investigation [50]. Smooth muscle is a major component also of other organs, such as intestine and urinary tract and indeed ASCs are already being used in experimental models of tissue engineering applied to urinary bladder [51]. ASCs represent a promising alternative for smooth muscle repair [52, 53].

ASCs not only have the potential to differentiate into cells and organs of mesodermal origin, but they are able to differentiate also into cells of nonmesodermal origin, such as epithelial cells [54] and neurons, which are of ectodermal origin, and endocrine pancreatic cells [55], hepatocytes [56], which are of endodermal origin. In particular, with a rather complex stimulation protocol, ASCs could be induced to express both neuronal and oligodendrocyte markers [57] and, when injected in ischemic rat brain, after being genetically engineered for the expression of the immortalizing gene telomerase or treated with azacytidine, they helped in functional recovery [58, 59]. ASCs have already been used to treat type I diabetes in experimental models [60, 61].

Recently, both murine and human ASCs have been reprogrammed to iPS (induced pluripotent stem) cells and shown to be even superior to conventional fibroblasts-derived iPS, since they can be obtained in feeder-free conditions [62]. This is a clear advantage, since it helps to establish animal-free, GMP-compliant systems for future therapeutic applications. These cells do not need feeder-layer cells because they produce a whole panel of growth factors contributing to their maintenance. Indeed the capability to produce and secrete bioactive molecules is another important characteristic of ASCs. These cells secrete significant amounts of growth/differentiation factors, such as VEGF, HGF, placental growth factor, FGF-2, TGF̃β-b, angiopoietin, LIF, activin A [62–66].

15.3.4 Immunomodulatory Properties of ASCs

Being wide-spread, available, accessible, easily harvestable and expandable in culture, ASCs represent a good option for autologous regenerative medicine, since they should not promote any rejection reaction when transplanted. Beside this, they have been reported to be endowed with anti-inflammatory power. As an example they are currently under evaluation in phase III clinical trials for inflammatory bowel disease and other intestinal disease manifestations [67] (see also Table 15.2). Indeed ASCs are able to decrease the release of a wide panel of inflammatory cytokines and chemokines and to increase the release of interleukin-10 levels [68]. Another and even more appealing property of these cells is the fact that they fail to stimulate an immune response, based on *in vitro* assays [69]. Indeed, after a few *in vitro* passages, ASCs lose or highly reduce the expression of many cell surface molecules involved in immune recognition. Not only are these cells not immunogenic, they are also immunoregulatory, since they can suppress immune responses, such as *in vitro* mixed lymphocyte reactions [69]. It is thus evident that ASC share with mesenchymal stem cells derived from bone marrow also this property, which, in the case of the bone marrow derived cells has been and is being exploited to prevent or control graft versus host reactions [70].

15.4 Adipose-Derived Stem Cells (ASCs) in Clinical Trials: The State of the Art

The cellular therapy based on the transplantation of adult stem cells opened new and exciting perspectives for the regeneration of damaged tissues. Today thousands of people all around the world are involved in clinical trials based on cell infusion/transplantation. Bone marrow-derived cells and blood circulating progenitors are the most used cell-sources up to now; however, in the last five years also ASCs have been classified as safe cell source by FDA (http://www.fda.gov) and the European Medicines Agency (http://www.ema.europa.eu) and started to be used in clinical trials in different countries as reported in Table 15.2.

As discussed above, cells obtained from the stroma of adipose tissue can be divided into minimally manipulated cell products, i.e., the heterogeneous cell population named Stromal Vascular Fraction (SVF) and the manipulated cells obtained after plastic adherence of SVF and named ASC [71].

Soft tissue replacement therapies, often required after tumor resection or trauma, are the more logical and simplest applications of adipose derived-cells [72, 73]. Breast reconstruction involved a great number of patients in a clinical study at University of Tokyo started 6 years ago, where more than 400 patients have been enrolled [71]. Breast augmentation is performed by the Cell-Assisted Lipotransfer (CAL), a technique characterized by the transplantation of aspirated fat containing fewer large vasculature structures. Matsumoto and collegues [74] demonstrated in an immune-deficient mouse model that human CAL transplantation is more effective in soft tissue augmentation compared to non-CAL transplantation characterized by vascular stromal fraction containing adipose stromal cells without vessels. More recently, the CAL transplantation, combined with SVF enriched in stromal progenitor cells, has been used in a small clinical trial of brest augmentation (15 patients). Clinical results after 12 months demonstrated a stable fat volume without necrosis or atrophy; these results suggest that progenitor-enriched CAL transplantation is a suitable methodology for the replacement of breast implants [75]. The CAL technology has been applied also for cosmetic surgery as demonstrated for facial lipoatrophy [76]. Another field of application for ASCs is related to the treatment of chronic open wounds such as decubit ulcers or wound and venous stasis ulcers in lower extremities in diabetic patients. A single blinded randomized clinical trial focusing on this problem started in 2008 at Washington DC Veteran Affairs Medical Center (clinicaltrials.gov identifier: NCT00815217).

Table 15.2 Clinical trials and case reports. Adapted from [2, 71, 86].

Indication	Study Type	Follow up	Reference
Soft Tissue			
Breast augmentation	Case Report	6 years	- 75
Craniofacial	Case Report	9-13 months	- 76
Orthopedic			
Calvaria	Case Report		- 77
Craniofacial	Case Report	12 months	- 78
Bone grafts using ASCs	Clinical Trial	Started in 2010	- clinicaltrials.gov identifier: NCT01218945
Immune Diseases			
Chron's Disease	Clinical Trial Phase I and Phase II	30 months 12 months	- 79, 80
Multiple Sclerosis	Case Report	12 months	- 81
Chronic Diseases			
ST-elevation Myocardial Infarction	Clinical Trial Phase I	Started in 2007	- Clinicaltrials.gov identifier: NCT00442806
Non revascularizable Ischemic Myocardum	Clinical Trial Phase I	Started in 2007	- Clinicaltrials.gov identifier: NCT00426868
Critical Limb Hischemia in diabetic patients	Clinical Trial	Started in 2010	- Clinicaltrials.gov identifier: NCT01257776
ASCs transplantation in Diabet type 2 patients	Clinical Trial	Started in 2008	- Clinicaltrials.gov identifier: NCT00703612

Autologous ASCs have been used successfully to treat a widespread traumatic calvarial bone defect for the first time in 2004 [77]. More recently they have been used for craniofacial bone reconstruction in a patient who underwent surgery for the removal of a cyst [78]. This case report paved the way for similar surgeries to other 20 patients. The repair of large osseous defects remains still an unsolved problem. For this reason a clinical trial was started at the University of Zürich in 2010 (clinicaltrials.gov identifier: NCT01218945) with the aim to evaluate the effectiveness of the implantation of pre-engineered synthetic large bone grafts loaded with ASCs; indeed the outcome of the use of ASCs in repairing critical defects in long bones still remains to be validated.

ASCs have been used also to threat immune diseases; in particular lesions and fistulas related to Chron's disease have been satisfactory treated with ASCs-based therapies [79–80].

ASCs have been also used to treat Multiple Sclerosis, an inflammatory disease characterized by degeneration of mielinated cells of CNS. In Costarica a Phase I clinical trial has started to evaluate the effectivnes of the treatment of patients affected from multiple sclerosis with ASCs [81].

Recently the possibility of treating chronic pathologies like neurodegenerative diseases, chronic heart failure and diabetes mellitus type 1 with ASCs-based therapy, which represent chronic diseases associated with high social and economic costs, has been explored. Different clinical trials were started in the last three years but the effectiveness of these treatments still remains to be confirmed with long-term analysis (see Table 15.2 for clinical trials details).

15.5 Conclusions and Future Directions

The use of adipose stem cells in humans is only beginning; nevertheless, great improvements have been made since the discovery of Zuk and colleagues in 2001 [8].

It has been discovered that the stromal compartment of adipose tissue contains mesenchymal cells similar for phenotype and features to those of bone marrow. ASCs have been deeply characterized *in vitro*, used in animal models and then their use for humans started to be approved.

However, essential questions are still waiting for a clear answer. The first question is "Are ASCs safe for human treatments?" Human adult mesenchymal stem cells have been used in several recent clinical trials, which reported their safety and their high resistance to transformation; different authors, however, described the similarities between adult normal stem cells and cancer stem cells. The possibility that SCs could undergo transformation has thus to be taken into account. The report from a Spanish research group from Madrid University describing that adult mesenchymal cells were spontaneously transformed upon long-term *in vitro* culture (>4–5 months) [82] has recently been retracted by the same researchers [83]. Apparently, murine SCs are more prone to *in vitro* transformation, especially if aged [84]. On these basis it is thus recommended to use ASCs within 3 months of *in vitro* expansion.

The second question is "Are ASCs effective tools for therapy?" Satisfactory results have been obtained in soft tissue replacement. Applications for adipose and bone tissue engineering are continuously growing in parallel with the new biomaterials and cell culture strategies. It has been demonstrated that ASCs improved the regeneration and the functional recovery for many different tissue *in vitro* and in animal models; nevertheless the mechanism involved is still unclear. In particular, it is unclear if the beneficial effect of ASCs is due to a direct differentiation of these immature progenitors into mature cells (ex. neuron, cardiomyocytes, endothelial cells) or mainly to a paracrine effect. Recent works suggest that the paracrine role is more important. Indeed it is well acknowledged that ASCs produce pro-survival molecules involved in neo-angiogenesis, modulation of the immune response and in the reduction of detrimental processes like inflammation and apoptosis.

In conclusion the "key" concept is the international standardization with the creation of standard procedures for cell management protocols, quality assurance control-testing criteria, together with the creation of specific ASCs Banks. Only this will help to clarify the ASCs potential and to understand their tumorigenesis-associated risk with transplantation.

References

[1] Barry, P. and Murphy, J.M. Mesenchymal stem cells: Clinical applications and biological characterization. *Int. J. Biochem. Cell Biol.*, 36(4):568–584, April 2004.
[2] Lindroos, B., Suuronen, R. and Miettinen, S. The potential of adipose stem cells in regenerative medicine. *Stem. Cell Rev.*, 7(2):269–291, June 2011.
[3] Takahashi, K. and Yamanaka, S. Induction of pluripotent stem cells from mouse embryonic and adult fibroblast cultures by defined factors. *Cell*, 126(4):663–676, August 2006.
[4] Young, F.E. A time for restraint. *Science*, 287(5457):1424, February 2000.
[5] Lenoir, N. Europe confronts the embryonic stem cell research challenge. *Science*, 287(5457):1425–1427, 2000.
[6] Pittenger, M.F., Mackay, A.M., Beck, S.C., Jaiswal, R.K., Douglas, R., Mosca, J.D., Moorman, M.A., Simonetti, D.W., Craig, S., Marshak and D.R. Multilineage potential of adult human mesenchymal stem cells. *Science*, 284(5411):143–147, April 1999.
[7] Friedenstein, A.J., Petrakova, K.V., Kurolesova, A.I. and Frolova, G.P. Heterotopic of bone marrow.Analysis of precursor cells for osteogenic and hematopoietic tissues. *Transplantation*, 6(2):230–247, March 1968.
[8] Zuk, P.A., Zhu, M., Mizuno, H., Huang, J., Futrell, J.W., Katz, A.J., Benhaim, P., Lorenz, H.P. and Hedrick, M.H. Multilineage cells from human adipose tissue: Implications for cell-based therapies. *Tissue Engineering*, 7(2):211–228, April 2001.
[9] Gimble, J.M., Katz, A.J. and Bunnell, B.A. Adipose-derived stem cells for regenerative medicine. *Circ. Res.*, 100(9),1249–1260, May 2007.

[10] Schä ffler, A. and Bü chler, C. Concise review: Adipose tissue-derived stromal cells – Basic and clinical implications for novel cell-based therapies. *Stem Cells*, 25(4):818–827, April 2007.

[11] Zuk, P.A. The adipose-derived stem cell: Looking back and looking ahead. *Mol. Biol. Cell*, 21(11):1783–1787, June 2010.

[12] Meza-Zepeda, L.A., Noer, A., Dahl, J.A., Micci, F., Myklebost, O. and Collas, P. High-resolution analysis of genetic stability of human adipose tissue stem cells cultured to senescence. *Journal of Cellular and Molecular Medicine* 12(2):553–563, April 2008.

[13] Yu, G., Wu, X., Dietrich, M.A., Polk, P., Scott, L.K., Ptitsyn, A.A. and Gimble, J.M. Yield and characterization of subcutaneous human adipose-derived stem cells by flow cytometric and adipogenic mRNA analyzes. *Cytotherapy*, 12(4):538–546, July 2010.

[14] Dominici, M., Le Blanc, K., Mueller, I., Slaper-Cortenbach, I., Marini, F., Krause, D., Deans, D., Keating, A., Prockop, Dj. and Horwitz, E. Minimal criteria for defining multipotent mesenchymal stromal cells. The International Society for Cellular Therapy position statement. *Cytotherapy*, 8(4):315–317, 2006.

[15] Horwitz, E.M., Le Blanc, K., Dominici, M., Mueller, I., Slaper-Cortenbach, I., Marini, F.C., Deans, R.J., Krause, D.S. and Keating, A. Clarification of the nomenclature for MSC: The International Society for Cellular Therapy position statement. *Cytotherapy*, 7(5):393–395, 2005.

[16] Gittenberger-de Groot, A.C., Winter, E.M. and Poelmann, R.E. Epicardium-derived cells (EPDCs) in development, cardiac disease and repair of ischemia. *J. Cell Mol. Med.*, 14(5):1056–1060, May 2010.

[17] Huang, S.C., Wu, T.C., Yu, H.C., Chen, M.R., Liu, C.M., Chiang, W.S. and Lin, K.M. Mechanical strain modulates age-related changes in the proliferation and differentiation of mouse adipose-derived stromal cells. *BMC Cell Biol.*, 11:18–31, March 2010.

[18] Folgiero, V., Migliano, E., Tedesco, M., Iacovelli, S., Bon, G., Torre, M.L., Sacchi, A., Marazzi, M., Bucher, S. and Falcioni, R. Purification and characterization of adipose-derived stem cells from patients with lipoaspirate transplant. *Cell Transplantation*, 19(10):1225–1235, 2010.

[19] Li, H., Fan, X., Kovi, R.C.. Jo, Y., Moquin, B., Konz, R., Stoicov, C., Kurt-Jones, E., Grossman, S.R., Lyle, S., Rogers, A.B., Montrose, M. and Houghton, J. Spontaneous expression of embryonic factors and p53 point mutations in aged mesenchymal stem cells: A model of age-related tumorigenesis in mice. *Cancer Research*, 67(22):10889–10898, November 2007.

[20] Lin, K., Matsubara, Y. and Masuda, Y. et al. Characterization of adipose tissue-derived cells isolated with the Celution system. *Cytotherapy*, 10:417–426, 2008.

[21] Cytori's Celution® 700 system to be regulated as a medical device by U.S. FDA. Available at: http://www.medicalnewstoday.com/articles/158091.php. July 2009.

[22] Tissue Genesis Cell Isolation System. Tissue Genesis Incorporated. Available at: http://www.tissuegenesis.com/. February 2009.

[23] Strem, B.M., Hicok, K.C., Zhu, M., Wulur, I., Alfonso, Z., Schreiber, R.E., Fraser, J.K. and Hedrick, M.H. Multipotential differentiation of adipose tissue-derived stem cells. *The Keio Journal of Medicine*, 54(3):132–141, September 2005.

[24] De Ugarte, D.A., Alfonso, Z. and Zuk, P.A. et al. Differential expression of stem cell mobilization-associated molecules on multi-lineage cells from adipose tissue and bone marrow. *Immunology Letters*, 89:267–270, 2003.

[25] Baglioni, S., Francalanci, M., Squecco, R., Lombardi, A., Cantini. G., Angeli, R., Gelmini, S., Guasti, D., Benvenuti, S., Annunziato, F., Bani, D., Liotta, F., Francini, F., Perigli, Serio, M. and Luconi, M. Characterization of human adult stem-cell populations isolated from visceral and subcutaneous adipose tissue. *FASEB J.*, 2(10):3494–3505, 2009.

[26] Mitchell, J.B., McIntosh, K., Zvonic, S., Garrett, S., Floyd, Z.E., Kloster, A., Di Halvorsen, Y., Storms, R.W., Goh, B., Kilroy, G., Wu, X. and Gimble, J.M. Immunophenotype of human adipose-derived cells: Temporal changes in stromal-associated and stem cell-associated markers. *Stem Cells*, 24(2):376–385, February 2006.

[27] Deslex, S., Negrel, R. and Vannier, C. et al. Differentiation of human adipocyte precursors in a chemically defined serum-free medium. *International Journal of Obesity*, 11:19–27, 1987.

[28] Hauner, H., Entenmann, G. and Wabitsch, M. et al. Promoting effect of glucocorticoids on the differentiation of human adipocyte precursor cells cultured in a chemically defined medium. *Journal of Clinical Investigation*, 84:1663–1670, 1989.

[29] Gregoire, F.M., Smas, C.M. and Sul, H.S. Understanding adipocyte differentiation. *Physiological Reviews*, 78:783–809, 1998.

[30] Aubin, J.E., Liu, F. and Malaval, L. et al. Osteoblast and chondroblast differentiation. *Bone*, 17:77S–83S, 1995.

[31] Halvorsen, Y.D., Franklin, D. and Bond, A.L. et al. Extracellular matrix mineralization and osteoblast gene expression by human adipose tissue-derived stromal cells. *Tissue Engineering*, 7:729–741, 2001.

[32] Wei, Y., Sun, X. and Wang, W. et al. Adipose-derived stem cells and chondrogenesis. *Cytotherapy*, 9:712–716, 2007.

[33] Koga, H., Engebretsen, L. and Brinchmann, J.E. et al. Mesenchymal stem cell-based therapy for cartilage repair: A review. *Knee Surgery, Sports Traumatology, Arthroscopy*, 2009.

[34] Liu, T.M., Martina, M. and Hutmacher, D.W. et al. Identification of common pathways mediating differentiation of bone marrow- and adipose tissue-derived human mesenchymal stem cells into three mesenchymal lineages. *Stem Cells*, 25:750–760, 2007.

[35] Planat-Benard, V., Menard, C., Andre, M., Puceat, M., Perez, A., Garcia-Verdugo, J.M., Penicaud, L. and Casteilla, L. Spontaneous cardiomyocyte differentiation from adipose tissue stroma cells. *Circulation Research*, 94(2):223–229, February 2004.

[36] van Dijk, A., Niessen, H.W. and Zandieh Doulabi, B. et al. Differentiation of human adipose-derived stem cells towards cardiomyocytes is facilitated by laminin. *Cell and Tissue Research*, 334:457–467, 2008.

[37] Valina, C., Pinkernell, K. and Song, Y.H. et al. Intracoronary administration of autologous adipose tissue-derived stem cells improves left ventricular function, perfusion, and remodelling after acute myocardial infarction. *European Heart Journal*, 28:2667–2677, 2007.

[38] Madonna, R., Geng, Y.J. and De Caterina, R. Adipose tissue-derived stem cells: characterization and potential for cardiovascular repair. *Arteriosclerosis, Thrombosis, and Vascular Biology*, 29:1723–1729, 2009.

[39] Miyahara, Y., Nagaya, N. and Kataoka, M. et al. Monolayered mesenchymal stem cells repair scarred myocardium after myocardial infarction. *Nat. Med.*, 12:459–465, 2006.

[40] Palpant, N.J. and Metzger, J.M. Aesthetic cardiology: Adipose-derived stem cells for myocardial repair. *Curr. Stem Cell Res. Ther.*, 5(2):145–152, June 2010.

[41] Jumabay M., Matsumoto, T., Yokoyama, S., Kano, K., Kusumi, Y., Masuko, T., Mitsumata, M., Saito, S., Hirayama, A., Mugishima, H. and Fukuda, N. Dedifferentiated fat cells convert to cardiomyocyte phenotype and repair infarcted cardiac tissue in rats. *J. Mol. Cell Cardiol.*, 47(5):565–575, November 2009.

[42] Di Rocco, G., Iachininoto, M.G. and Tritarelli, A. et al. Myogenic potential of adipose-tissue-derived cells. *J. Cell Sci.*, 119:2945–2952, 2006.

[43] Lee, J.H. and Kemp, D.M. Human adipose-derived stem cells display myogenic potential and perturbed function in hypoxic conditions. *Biochem. Biophys. Res. Commun.*, 341:882–888, 2006.

[44] Planat-Benard, V., Silvestre, J.S. and Cousin, B. et al. Plasticity of human adipose lineage cells toward endothelial cells: Physiological and therapeutic perspectives. *Circulation*, 109:656–663, 2004.

[45] Miranville, A., Heeschen, C. and Sengenes, C. et al. Improvement of postnatal neovascularization by human adipose tissue-derived stem cells. *Circulation*, 110:349–355, 2004.

[46] De Francesco, F., Tirino, V., Desiderio, V., Ferraro, G., D'Andrea, F., Giuliano, M., Libondi, G., Pirozzi, G., De Rosa, A. and Papaccio, G. Human CD34/CD90 ASCs are capable of growing as sphere clusters, producing high levels of VEGF and forming capillaries. *PLoS One*, 6(4(8)):e6537, August 2009.

[47] Moon, M.H., Kim, S.Y. and Kim, Y.J. et al. Human adipose tissue-derived mesenchymal stem cells improve postnatal neovascularization in a mouse model of hindlimb ischemi. *Cell Physiol. Biochem.*, 17:279–290, 2006.

[48] Rubina, K., Kalinina, N., Efimenko, A., Lopatina, T., Melikhova, V., Tsokolaeva, Z., Sysoeva, V., Tkachuk, V. and Parfyonova, Y. Adipose stromal cells stimulate angiogenesis via promoting progenitor cell differentiation, secretion of angiogenic factors, and enhancing vessel maturation. *Tissue Eng. Part A*, 15(8):2039–2050, August 2009.

[49] Merfeld-Clauss, S., Gollahalli, N., March, K.L. and Traktuev, D.O. Adipose tissue progenitor cells directly interact with endothelial cells to induce vascular network formation. *Tissue Eng. Part A*, 16(9):2953–2966, September 2010.

[50] Wang, C., Cen, L., Yin, S., Liu, Q., Liu, W., Cao, Y. and Cui, L. A small diameter elastic blood vessel wall prepared under pulsatile conditions from polyglycolic acid mesh and smooth muscle cells differentiated from adipose-derived stem cells. *Biomaterials*, 31(4):621–630, February 2010.

[51] Jack, G.S., Zhang, R., Lee, M., Xu, Y., Wu, B.M. and Rodrìguez, L.V. Urinary bladder smooth muscle engineered from adipose stem cells and a three dimensional synthetic composite. *Biomaterials*, 30(19):3259–3270, July 2009.

[52] Rodrìguez, L.V., Alfonso, Z., Zhang, R., Leung, J., Wu, B. and Ignarro, L.J. Clonogenic multipotent stem cells in human adipose tissue differentiate into functional smooth muscle cells. *Proc. Natl. Acad. Sci. USA*, 103(32):12167–12172, August 2008.

[53] de Villiers, J.A., Houreld, N. and Abrahamse, H. Adipose derived stem cells and smooth muscle cells: Implications for regenerative medicine. *Stem Cell Rev.*, 5(3):256–265, September 2009.

[54] Brzoska, M., Geiger, H., Gauer, S. and Baer, P. Epithelial differentiation of human adipose tissue-derived adult stem cells. *Biochem. Biophys. Res. Commun.*, 330(1):142–150, April 2005.

[55] Timper, K., Seboek, D. and Eberhardt M. et al. Human adipose tissue-derived mesenchymal stem cells differentiate into insulin, somatostatin, and glucagon expressing cells. *Biochem. Biophys. Res. Commun.*, 341:1135–1140, 2006.

[56] Seo, M.J., Suh, S.Y. and Bae, Y.C. et al. Differentiation of human adipose stromal cells into hepatic lineage in vitro and in vivo. *Biochem. Biophys. Res. Commun.*, 328:258–264, 2005.

[57] Ning, H., Lin, G., Lue, T.F. and Lin, C.S. Neuron-like differentiation of adipose tissue-derived stromal cells and vascular smooth muscle cells. *Differentiation*, 74(9–10):510–518, December 2006.

[58] Jun, E.S., Lee, T.H. and Cho, H.H. et al. Expression of telomerase extends longevity and enhances differentiation in human adipose tissue-derived stromal cells. *Cell Physiol. Biochem.*, 14:261–268, 2004.

[59] Kang, S.K., Lee, D.H. and Bae, Y.C. et al. Improvement of neurological deficits by intracerebral transplantation of human adipose tissue-derived stromal cells after cerebral ischemia in rats. *Exp. Neurol.*, 183:355–366, 2003.

[60] Kang, H.M., Kim, J. and Park, S., et al. Insulin-secreting cells from human eyelid-derived stem cells alleviate type I diabetes in immunocompetent mice. *Stem Cells*, 27:1999–2008, 2009.

[61] Lin, G., Wang., G. and Liu, G. et al. Treatment of type 1 diabetes with adipose tissue-derived stem cells expressing pancreatic duodenal homeobox 1. *Stem Cells and Development*, 18:1399–1406, 2009.

[62] Sugii, S., Kida, Y., Kawamura, T., Suzuki, J., Vassena, R., Yin, Y.Q., Lutz, M.K., Berggren, W.T., Izpisú a, J.C. Belmonte and Evans, R.M. Human and mouse adipose-derived cells support feeder-independent induction of pluripotent stem cells. *Proc. Natl. Acad. Sci. USA*, 107(8):3558–3563, February 2010.

[63] Nakagami, H., Morishita, R. and Maeda K. et al. Adipose tissue-derived stromal cells as a novel option for regenerative cell therapy. *J. Atheroscler. Thromb.*, 13:77–81, 2006.

[64] Nakagami, H., Maeda, K. and Morishita, R. et al. Novel autologous cell therapy in ischemic limb disease through growth factor secretion by cultured adipose tissue-derived stromal cells. *Arterioscler. Thromb. Vasc. Biol.*, 25:2542–2547, 2005.

[65] Rehman, J., Traktuev, D. and Li, J. et al. Secretion of angiogenic and antiapoptotic factors by human adipose stromal cells. *Circulation*, 109:1292–1298, 2004.

[66] Cao, Y., Sun, Z. and Liao, L. et al. Human adipose tissue-derived stem cells differentiate into endothelial cells in vitro and improve postnatal neovascularization in vivo. *Biochem. Biophys. Res. Commun.*, 332:370–379, 2005.

[67] Lanzoni, G., Alviano, F., Marchionni, C., Bonsi, L., Costa, R., Foroni, L., Roda, G., Belluzzi, A., Caponi, A., Ricci, F., Tazzari, P.L., Pagliaro, P., Rizzo, R., Lanza, F., Baricordi, O.R.. Pasquinelli, G., Roda, E. and Bagnara, G.P. Isolation of stem cell populations with trophic and immunoregulatory functions from human intestinal tissues: Potential for cell therapy in inflammatory bowel disease. *Cytotherapy*, 11(8):1020–1031, 2009.

[68] González, M.A., Gonzalez-Rey, E., Rico, L., Büscher, D. and Delgado, M. Adipose-derived mesenchymal stem cells alleviate experimental colitis by inhibiting inflammatory and autoimmune responses. *Gastroenterology*, 136(3):978–989, March 2009.

[69] McIntosh, K., Zvonic, S., Garrett, S., Mitchell, J.B., Floyd, Z.E.,Hammill, L., Kloster, A., Di Halvorsen, Y., Ting, J.P., Storms, R.W., Goh, B., Kilroy, G., Wu, X. and Gimble, J.M. The immunogenicity of human adipose-derived cells: Temporal changes in vitro. *Stem Cells*, 24(5):1246–1253, May 2006.

[70] Fang, B., Song, Y.P., Liao, L.M.. Han, Q. and Zhao, R.C. Treatment of severe therapy-resistant acute graft-versus-host disease with human adipose tissue-derived mesenchymal stem cells. *Bone Marrow Transplant*, 38(5):389–390, September 2006.

[71] Gimble, J.M., Guilak, F. and Bunnell, B.A. Clinical and preclinical translation of cell-based therapies using adipose tissue-derived cells. *Stem Cell Research and Therapy*, 1(2):19, 29 June 2010.

[72] Patrick, C.W., Jr. Tissue engineering strategies for adipose tissue repair. *The Anatomical Record*, 263(4):361–366, August 2001.

[73] Yoshimura, K., Sato, K., Aoi, N., Kurita, M., Hirohi, T. and Harii, K. Cell-assisted lipotransfer for cosmetic breast augmentation: Supportive use of adipose-derived stem/stromal cells. *Aesthetic Plast. Surg.*, 32(1):48–57, January 2008.

[74] Matsumoto, D., Sato, K., Gonda, K., Takaki, Y., Shigeura, T., Sato, T., Aiba-Kojima, E., Iizuka, F., Inoue, K., Suga, H. and Yoshimura, K. Cell-assisted lipotransfer: Supportive use of human adipose-derived cells for soft tissue augmentation with lipoinjection. *Tissue Engineering*, 12(12):3375–3382, December 2006.

[75] Yoshimura, K., Asano, Y., Aoi, N., Kurita, M., Oshima, Y., Sato, K., Inoue, K., Suga, H., Eto, H., Kato, H. and Harii, K. Progenitor-enriched adipose tissue transplantation as rescue for breast implant complications. *Breast Journal*, 16(2):169–175, March–April 2010.

[76] Yoshimura, K., Sato, K., Aoi, N., Kurita, M., Inoue, K., Suga, H., Eto, H., Kato, H., Hirohi, T. and Harii, K. Cell-assisted lipotransfer for facial lipoatrophy: Efficacy of clinical use of adipose-derived stem cells. *Dermatological Surgery*, 34(9):1178–1185, September 2008.

[77] Lendeckel, S., Jödicke, A., Christophis, P., Heidinger, K., Wolff, J., Fraser, J.K., Hedrick, M.H., Berthold, L. and Howaldt, H.-P. Autologous stem cells (adipose) and fibrin glue used to treat widespread traumatic calvarial defects: Case report. *Journal of Cranio-Maxillofacial Surgery*, 32(6):370–373, December 2004.

[78] Mesimaki, K., Lindroos, B., Tornwall, J., Mauno, J., Lindqvist, C., Kontio, R., Miettinenv, S. and Suuronen, R. Novel maxillary reconstruction with ectopic bone formation by GMP adipose stem cells. *Int. J. Oral Maxillofac. Surg.*, 38(3):201–209, March 2009.

[79] García-Olmo, D., García-Arranz, M., Herreros, D., Pascual, I., Peiro, C. and Rodríguez-Montes, J.A. A phase I clinical trial of the treatment of Crohn's fistula by adipose mesenchymal stem cell transplantation. *Dis. Colon Rectum*, 48(7):1416–1423, July 2005.

[80] García-Olmo, D., Herreros, D., Pascual, I., Pascual, J.A., Del-Valle, E., Zorrilla, J., De-La-Quintana, P., García-Arranz, M. and Pascual M. Expanded adipose-derived stem cells for the treatment of complex perianal fistula: A phase II clinical trial. *Dis. Colon Rectum*, 52(1):79–86, January 2009.

[81] Riordan, N.H., Ichim, T.E., Min, W.P., Wang, H., Solano, F., Lara, F., Alfaro, M., Rodriguez, J.P., Harman, R.J., Patel, A.N., Murphy, M.P., Lee, R.R. and Minev, B. Non-expanded adipose stromal vascular fraction cell therapy for multiple sclerosis. *J. Translational Medicine*, 24(7):29, April 2009.

[82] Rubio, D., Garcia-Castro, J., Martín, M.C., de la Fuente, R., Cigudosa, J.C., Lloyd, A.C. and Bernad, A. Spontaneous human adult stem cell transformation. *Cancer Research*, 65(8):3035–3059, April 2005. Erratum in: *Cancer Res.* 65(11):4969, 1 June 2005.

[83] de la Fuente, R., Bernad, A., Garcia-Castro, J., Martin, M.C. and Cigudosa J.C. *Cancer Res.*, 70(16):6682, August 2010.

[84] Li, H., Fan, X., Kovi, R.C., Jo, Y.J., Moquin, B., Konz, R., Stoicov, C., Kurt-Jones, E., Grossman, S.R., Lyle, S., Rogers, A.B., Montrose, M. and Houghton, J.M. Spontaneous expression of embryonic factors and p53 point mutations in aged mesenchymal stem cells: A model of age-related tumorigenesis in mice. *Cancer Research*, 67(22):10889–10898, November 2007.

[85] Kern, S., Eichler, H., Stoeve, J., Klü ter, H. and Bieback, K. Comparative analysis of mesenchymal stem cells from bone marrow, umbilical cord blood, or adipose tissue. *Stem Cells*, 24(5):1294–1301, May 2006.

[86] Mizuno H. Adipose-derived stem cells for tissue repair and regeneration: Ten years of research and a literature review. *J. Nippon Med. Sch.*, 76:56–66, 2009.

16

Embryonic Stem Cells in Cardiac Regeneration

Dinender K. Singla

Biomolecular Science Center, University of Central Florida, Orlando, FL, USA;
e-mail: dsingla@mail.ucf.edu

Abstract

Following myocardial infarction, the presence of cell loss due to apoptosis as well as fibrosis leading to cardiac remodeling which ultimately hardens the heart muscle and causes cardiac dysfunction has been confirmed. Despite significant improvement in therapeutic options, cardiovascular diseases are still on the rise in both developed and developing countries. Cell therapy has gained significant attention following successful, confirmed data using various adult and embryonic stem cells in experimental studies. This data spurred clinical trials which show exciting beneficial effects at the short term follow-up. However, these effects were not statistically significant when observed at 18 months follow up. Unfortunately, with faded hope, these despairing results sent researchers back to the bench to identify another optimal cell type for cardiac repair and regeneration. Following extensive research, we now believe that embryonic stem (ES) cells have better potential to differentiate into cardiac myocytes compared with counterpart adult stem cells. In this review chapter, we focused our attention on ES cells and their potential in cardiac regeneration.

Keywords: myocardial infarction, heart regeneration, stem cells, teratoma, cardiac remodeling.

P. Di Nardo (Ed.), Adult Stem Cell Standardization, 209–221.

16.1 Introduction

Cardiovascular diseases (CVDs) are a major health problem in developed countries as well as in rapidly developing countries and the incidence of CVDs is steadily increasing [30]. This significant increase in CVDs has also substantially increased mortality and morbidity rates [30]. Moreover, the escalating incidence of CVD suggests quality of life is deteriorating day by day in the young and old population. Therefore, there is an urgent need to reevaluate the current medications used to treat patients as well as develop new therapeutic interventions. The current therapeutic options for the treatment of cardiac diseases include use of statins, angiotensin converting enzyme inhibitors, carvedilol and various antioxidant and beta-blocker drugs [5, 10, 48, 59]. Well documented, CVDs are associated with significant cell loss due to cell death via apoptosis and necrosis [2, 26]. However, current therapeutic strategies do not possess the potential to regenerate the injured myocardium necessary for enhanced heart contractility. Moreover, patients with end stage heart failure have very limited options including waiting for a heart transplantation. However, recent statistics indicate the number of patients waiting for a donor heart for transplantation is significantly higher than the number of actual available hearts [30]. For this scenario, the current hope is that recent successful transplantations of various stem cells in animal models will become a reality for clinical practice in the very near future.

Stem cell transplantation studies have shown the differentiation of cardiac myocytes, a specialized cell in the heart responsible for heart contractility, as well as vascular smooth muscle and endothelial cells, required to regenerate coronary blood vessels for adequate blood supply in the heart [15, 50]. Formation of these cardiac cell types in the heart were well documented when bone marrow stem cells, cardiac stem cells, and embryonic stem (ES) cells were transplanted in the injured myocardium [6, 20, 50]. However, we are still learning to identify the most suitable donor stem cell type required for heart regeneration.

16.2 Cell Types Demonstrating Their Potential for Heart Regeneration

The first cell types examined for cardiac repair and regeneration were skeletal myoblasts transplanted in animals and used in human clinical studies [11, 12, 17, 32]. Thereafter, neonatal cardiac myocytes were also examined to determine whether they could regenerate injured myocardium following transplantation [11]. Most striking was an adult stem cell transplantation study in 2001 performed by Dr. Anversa's group which showed that transplanted bone marrow (BM) stem cells regenerated infarcted mouse myocardium [41]. Subsequently, many groups have examined various cell types such as smooth muscle cells, fetal and adult cardiomyocytes, BM stromal and hematopoietic stem cells, and mouse ES cells [23, 32, 33, 50]. Recently, studies have shown generated iPS cells from mouse, rat (Singla et al., unpublished data) and human somatic cells have the potential to regenerate infarcted myocardium following transplantation [37, 38]. Moreover, in addition to differentiation into cardiac cell types and regeneration of injured myocardium, most of the cell transplantation studies demonstrated significant improved cardiac function in the infarcted heart, ischemia-reperfusion injury and dilated cardiomyopathy [7, 38, 50].

Next, cardiac stem cells (CSCs), a unique cardiac specific stem cell population present in the adult heart of mice, rat and humans was also discovered by Dr. Anversa's group [6, 35, 64]. These cells have been shown to be undifferentiated, self-renewing, clonogenic and have the capability to regenerate the heart by differentiating into all three major heart cell types [35]. Since resident CSCs are already present in the adult heart, identification of appropriate growth factors that translocate these CSCs from their storage niches to the injury site as well as differentiate into heart cell types required for regeneration is eminent. Our recently published data demonstrates identification of c-kit^{+ve} CSCs and their activation as well as differentiation into neovascularization in the infarcted heart upon transplantation of ES cell conditioned medium (CM) [53]. Moreover, genetic modifications with the anti-apoptotic gene Akt in cardiomyocytes and mesenchymal stem cells (MSCs) have demonstrated an increase in engraftment, and a decrease in apoptosis along with improved cardiac function in the infarcted heart [39]. Human and mouse ES cells have also provided significant improvement in cardiac regeneration in the models of mouse, rat and sheep infarcted hearts [18, 31, 51]. Overall, these successful laboratory studies have provided intriguing interest in stem cell research. However, cell transplantation studies were unable to regenerate the injured myocardium enough to reestablish adequate normal blood supply.

16.3 Adult Stem Cells in the Clinic

Studies exploiting transplantation of BM stem cells and their sub populations, including CSCs and MSCs, into various mouse, rat and pig models have demonstrated successful cardiac regeneration along with improved cardiac geometry and function [20, 23]. This basic research has raised hope very high and forced various clinician scientists to examine the potential use of BM stem cells and their sub-populations as well as other adult stem cells in patients. In this regard, adult stem cells were transplanted in patients with coronary artery disease and their ability to improve patients' life style and cardiac function was examined in a follow-up from day 0 through 18 months [19]. More precisely, acute myocardial infarcted (AMI) patients were identified, their consent was taken and bone marrow (BM) cells were isolated for the derivation of progenitor cells [46]. Afterwards, BM derived progenitor cells were expanded to the desired cell number ($5.5 \pm 3.9 \times 10^6$) required for transplantation [45, 47]. Using intracoronary infusion, REPAIR-AMI patients were transplanted with autologous BM-derived progenitor cells. Up to 12 months post-transplantation, left ventricular ejection fraction (LVEF) was examined and results demonstrated an increase in cardiac function [47]. Moreover, this was an early clinical trial performed on patients to determine safety and feasibility of transplanted BMS cells in patients. The major limitation of the REPAIAR-AMI study of patient data following stem cell transplantation, was that they were not analyzed in a randomized fashion. This limitation was addressed in the randomized BOOST trials [63]. In this study, autologous BM-derived mononuclear cells were transplanted in the infarcted heart patients and demonstrated improved LVEF at short term (6 months) follow-up [63]. Noticeably, the BOOST clinical trial was extended to examine their effects on long term cardiac function. However, increased enthusiasm from the short term study (6 months) disappeared as observed cardiac function at 18 months showed no significant improvement. A large clinical trial named the REGENT trial was conducted to examine the potential of BM subpopulations such as BM-mononulcear cells versus selected $CD34^+CXCR4^+$ BM cells to improve cardiac function in 200 AMI patients [60]. Results obtained from this randomized study showed no significant improvement in LVEF for either treatment group compared with controls after 6 months [60]. Moreover, other parameters of the analyzed findings including death, stroke, neovascularization and re-infarction also demonstrated no statistical difference between the groups [60]. To date, most of the clinical studies tested BM progenitor and mononuclear cells, and BM specific $CD34^+CXCR4^+$ cells to treat injured myocardium [47, 60, 63]. However, there is no clinical study that provides convincing evidence of the beneficial effects in cardiac repair and regeneration in the long-term follow-up. Clinical trials using CSCs and MSCs in the United States are in progress and early findings are strikingly positive. However, solid conclusions have not yet been reached.

16.4 ES Cells in the Cell Culture System and the Heart

Following fertilization, generated totipotent oocytes cleave into 2, 4 and 8 cell stages respectively leading to the preimplantation blastocyst containing the inner cell mass (ICM). The ICM gives rise to functional life which primarily consists of all three embryonic germ layers; ectoderm, endoderm and meso-derm. Noticeably, ES cells are isolated from ICM. Therefore, to establish ES cell lines in the cell culture, ES cells must be able to propagate without differ-entiation in cell culture as well as preserve their characteristic to differentiate into all three embryonic germ layers. Currently, ES cells have been isolated from mouse, rat, monkey, pig and human and provide well established cell lines for future research and therapeutic use as needed [8, 14, 62]. Also, in the cell culture system, formation of many differentiated cell types generated from all three germ layers by ES cells has been achieved [61]. To achieve differentiation of various cell types in the cell culture system, scientists have developed a 3D method involving cell aggregates called embryoid bodies (EBs). Studies have shown these EBs stain positive for differentiated cell types of the heart including cardiac-α-action for cardiac myocytes, $CD31^{+ve}$ for endothelial cells, and smooth muscle α-actin for smooth muscle cells [50]. Similarly, from EBs, differentiated cells including neurons to represent brain cell types, proximal cells for kidney and hepatocytes, to determine presence of liver cells, were identified by numerous investigators [25, 36, 44, 61]. Human and mouse ES cells have been shown to be the best stem cells to use to examine organ specific cell type differentiation compared to rat and pig ES cells. Importantly, numerous studies have shown proof of the prin-ciple demonstrating cardiac, brain, kidney and liver cell generation in the cell culture system [25, 36, 44, 61]. This data proved very exciting and raised hopes of the possibilities of transplanting generated cells in various organs as required during disease development. However, the number of tissue specific differentiated cells determined in the cell culture system remains significantly low compared with the amount of cells required for complete regeneration of injured organs such as the heart [56]. Current research has been focused on the development of novel approaches to increase cell type differenti-ation from both human and mouse ES cells. These strategies include genetic modification of ES cells as well as priming the cells with various growth factors/cytokines. For example; we showed that treatment of mouse ES cells with TGF-β2 (required for embryonic cardiogenesis) significantly enhanced spontaneous and rhythmically beating cardiomyocytes [54]. Interestingly, use of bone morphogenic proteins (BMP2) and vitamin A and C treatment of ES cells have also demonstrated increased differentiation of beating cardiac myocytes [29, 34, 42]. Additionally, our unpublished data shows fibroblast growth factor (FGF) 8 enhances cardiac myocyte differentiation in the cell culture system.

After exciting cell culture data, many investigators transplanted ES cells in various diseased organs including brain, kidney, liver and heart. Here, we will focus on heart regeneration following ES cell transplantation. Transplanted ES cells in the infarcted mouse, rat and sheep hearts demonstrated new cardiac myocyte cell type differentiation along with the formation of gap junction protein connexin-43 up to 12 weeks post-MI [25, 31, 50]. Moreover, transplanted ES cells also demonstrated improved cardiac function [51]. However, it remains unclear how much actual engraftment into the myocardium and differentiation into cardiomyocytes takes place relative to the number of transplanted ES cells [27, 40]. Moreover, there is no study which shows the true amount of cell engraftment versus differentiation per mm^2 heart regeneration. It was widely accepted by various investigators until now that the amount of regeneration required for normal heart functioning is still a long way off. Therefore, general consensus has reached the point that improved cell transplantation strategies are needed to enhance engraftment and differentiation.

However, among the various options to enhance cardiac regeneration in the infarcted heart, the most frequent attempt was to use genetically modified stem cells or identification of pro-cardiac growth factors to induce differentiation into cardiac myocytes following transplantation of growth specific primed ES cells. In this regard, insulin growth factor-1 primed ES cells were transplanted into the injured myocardium and demonstrated enhanced cardiac myocyte differentiation as well as significantly improved cardiac function compared with wild type ES cells [13]. Similarly, growth factors including TGF-β and FGF demonstrated their capability in the generation of beating cardiac myocytes in cell culture, furthermore, showed if ES cells were primed with either of these two growth factors, then transplanted into infarcted mouse hearts they would demonstrate a significant increase in the donor cell engraftment, cardiac myocyte differentiation as well as formation of gap junction proteins as confirmed with connexin-43 [31, 43, 54]. Primed ES cells in these studies further confirmed enhanced improved cardiac function compared to unprimed ES cells [31, 54].

16.5 Teratoma Formation Following ES Cell Transplantation

Human and mouse ES cells form complex teratomas in immune-deficient mice which is a well-recognized characteristic of these cells [27, 40, 49]. The formation of complex teratoma following transplantation has been recognized as a major limitation of ES cells [49]. Therefore, clinical use of these cells has been recognized as limited as long as approved extra care is taken. Recently, a US human stem cell company, Geron, was granted permission to conduct a clinical trial using human ES cells in patients with new traumatic spinal cord injuries.

Various experimental studies have demonstrated that undifferentiated ES cells either do or do not form teratoma in the infarcted heart [9, 27, 57]. This has been resolved recently after carefully examining the dose of stem cell transplantation [27, 40]. Teratoma formation in the infracted heart was examined by Nussbaum et al. by using varying doses of ES cells and determined that transplantation of up to 100,000 undifferentiated ES cells in the mouse heart was safe as no teratoma was observed [40]. This study corroborated our published data in which we transplanted 30,000 undifferentiated ES cells and determined there was no presence of teratoma formation [51]. Moreover, formation of teratoma was evident if more than 100,000 ES cells are injected into non infarcted or infarcted mouse hearts [40]. In contrast, no formation of teratoma was observed when there was transplantation of up to 300,000 undifferentiated ES cells into the infarcted heart. This study was examined for the period of 12 weeks [21]. The exact reason for this discrepancy for the formation of teratoma following transplantation of 100,000 cells compared with no teratoma with 300,000 is not well defined. Therefore further studies are required to understand the clear mechanisms of teratoma formation in the infarcted heart.

16.6 Cardiac Remodeling Following ES Cell Transplantation

Various pathological mechanisms including oxidative stress induce cardiac myocyte cell loss in the infarcted heart [24, 26]. Depletion of cardiac myocyte, endothelial and vascular smooth muscle cells generates scar tissue which is replaced with proliferating cardiac fibroblasts that give rise to alterations in cardiac geometry as well as rearrangement of diseased myocardium. This ultimately hardens the tissue and decreases cardiac function [22, 58].

Moreover, cardiac remodeling also involves infiltration of macrophages, neutrophils and mast cells (inflammatory cells). Next, the degradation of extracellular matrix (ECM) via activation of matrix metalloproteinases (MMPs) was observed in the infarcted hearts.

Apoptosis, or programmed cell death, is a highly orchestrated biochemical process characterized by cellular changes including blebbing and shrinkage, nuclear condensation and DNA fragmentation. Collective evidence suggests apoptosis plays a vital role in not only normal heart development but also in the pathogenesis of various cardiovascular diseases including MI [16, 26] Following MI, cardiac myocyte apoptosis is initiated by a host of stimuli including cytokines, reactive oxygen species (ROS), and DNA damage [4, 24, 26]. Cardiac myocyte cell death via apoptosis consequent to MI leads to a host of pathophysiological changes including fibrosis, hypertrophy, cardiac dysfunction, and heart failure [2, 3, 26]. Countless research efforts have gone into attenuating post-MI myocardial apoptosis through the use of diverse stem cell populations including ES cells and their generated conditioned media (CM) [1, 28, 51, 54, 55]. Previous *in vitro* studies have demonstrated apoptosis, assessed in H9c2 cardiomyoblasts, was blunted by cytoprotective released factors from ES cells and this anti-apoptotic effect was mediated through the PI3K/Akt pathway [52, 55]. Moreover, CM generated from ES cells primed with TGF-β2 further inhibited iodoacetic acid (IAA) and H_2O_2 induced apoptosis in the cell culture model [54]. Most notably, *in vivo* studies have revealed that transplanted ES cells or their CM inhibited cardiac and vascular cell apoptosis in the infarcted myocardium paralleled with improved cardiac function [51, 53, 54]. Our recently published data suggests that anti-apoptotic factors released from ES cells inhibits apoptosis and fibrosis in the infarcted heart [52, 53].

16.7 Future Directions

Multiple stem cells have been examined in cardiovascular diseases to inhibit apoptosis, fibrosis as well as generate new heart cell types for better cardiac function. These studies have provided significant improvement in cardiac function along with the inhibition of apoptosis and cardiac remodeling. The significant failure of adult stem cells in clinical trials has raised many questions such as: (1) What is the optimal cell type? (2) Can we increase or decrease the dose of transplanted stem cells for better cardiac function in the long term? (3) Can genetically modified stem cells enhance cardiac regeneration? (4) What is the exact role of secreted autocrine or paracrine factors in

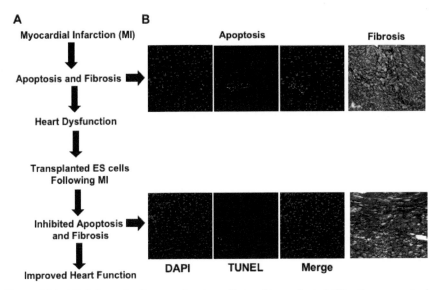

Figure 16.1 (A) Schematic diagram showing effects of transplanted ES cells on apoptosis, fibrosis and heart function following myocardial infarction in the heart. (B) shows representative photomicrographs of apoptosis; DAPI stained nuclei in blue; TUNEL stained apoptotic nuclei appeared in red; MERGE; shows DAPI+TUNEL stained merged nuclei. The right panel shows sections stained with Masson's staining; blue area shows total fibrosis. Data is from the mouse hearts with and without ES cells transplantation at D14 post-MI.

cardiac repair? (5) How many secreted factors truly inhibit apoptosis versus fibrosis versus enhancing endogenous cardiac regeneration? Overall, the stem cell therapy field has advanced significantly but more refined strategies are needed to achieve better understanding.

References

[1] Angoulvant, D., Ivanes, F., Ferrera, R., Matthews, P.G., Nataf, S. and Ovize, M. Mesenchymal stem cell conditioned media attenuates in vitro and ex vivo myocardial reperfusion injury. *J. Heart Lung Transplant.*, 2010.

[2] Anversa, P. and Kajstura, J. Myocyte cell death in the diseased heart. *Circ. Res.*, 82:1231–1233, 1998.

[3] Anversa, P., Kajstura, J. and Olivetti, G. Myocyte death in heart failure. *Curr. Opin. Cardiol.*, 11:245–251, 1996.

[4] Anversa, P., Olivetti, G., Leri, A., Liu, Y. and Kajstura, J. Myocyte cell death and ventricular remodeling. *Curr. Opin. Nephrol. Hypertens.*, 6:169–176, 1997.

[5] Basile, J.N. One size does not fit all: The role of vasodilating beta-blockers in controlling hypertension as a means of reducing cardiovascular and stroke risk. *Am. J. Med.*, 123:S9–S15, 2010.

[6] Beltrami, A.P., Barlucchi, L., Torella, D., Baker, M., Limana, F., Chimenti, S., Kasahara, H., Rota, M., Musso, E., Urbanek, K., Leri, A., Kajstura, J., Nadal-Ginard, B. and Anversa, P. Adult cardiac stem cells are multipotent and support myocardial regeneration. *Cell*, 114:763–776, 2003.

[7] Boudoulas, K.D. and Hatzopoulos, A.K. Cardiac repair and regeneration: The Rubik's cube of cell therapy for heart disease. *Dis. Model Mech.*, 2:344–358, 2009.

[8] Brevini, T.A., Antonini, S., Cillo, F., Crestan, M. and Gandolfi, F. Porcine embryonic stem cells: Facts, challenges and hopes. *Theriogenology*, 68(Suppl 1):S206–S213, 2007.

[9] Cao, F., Lin, S., Xie, X., Ray, P., Patel, M., Zhang, X., Drukker, M., Dylla, S.J., Connolly, A.J., Chen, X., Weissman, I.L., Gambhir, S.S. and Wu, J.C. In vivo visualization of embryonic stem cell survival, proliferation, and migration after cardiac delivery. *Circulation*, 113:1005–1014, 2006.

[10] Das, D.K. and Maulik, N. Resveratrol in cardioprotection: A therapeutic promise of alternative medicine. *Mol. Interv.*, 6:36–47, 2006.

[11] Dowell, J.D., Rubart, M., Pasumarthi, K.B., Soonpaa, M.H. and Field, L.J. Myocyte and myogenic stem cell transplantation in the heart. *Cardiovasc. Res.*, 58:336–350, 2003.

[12] Durrani, S., Konoplyannikov, M., Ashraf, M. and Haider, K.H. Skeletal myoblasts for cardiac repair. *Regen. Med.*, 5:919–932, 2010.

[13] Enoki, C., Otani, H., Sato, D., Okada, T., Hattori, R. and Imamura, H. Enhanced mesenchymal cell engraftment by IGF-1 improves left ventricular function in rats undergoing myocardial infarction. *Int. J. Cardiol.*, 138:9–18, 2010.

[14] Evans, M.J. and Kaufman, M.H. Establishment in culture of pluripotential cells from mouse embryos. *Nature*, 292:154–156, 1981.

[15] Fatma, S., Selby, D.E., Singla, R.D. and Singla, D.K. Factors released from embryonic stem cells stimulate c-kit-FLK-1(+ve) progenitor cells and enhance neovascularization. *Antioxid. Redox. Signal*, 13:1857–1865, 2010.

[16] Fisher, S.A., Langille, B.L. and Srivastava, D. Apoptosis during cardiovascular development. *Circ. Res.* 87:856–864, 2000.

[17] Forte, E., Chimenti, I., Barile, L., Gaetani, R., Angelini, F., Ionta, V., Messina, E. and Giacomello, A. Cardiac Cell Therapy: The Next (Re)Generation. *Stem Cell Rev.*, 2011.

[18] Fukuda, K. and Yuasa, S. Stem cells as a source of regenerative cardiomyocytes. *Circ. Res.*, 98:1002-1013, 2006.

[19] Haider, H.K. and Ashraf, M. Bone marrow cell transplantation in clinical perspective. *J. Mol. Cell Cardiol.*, 38:225–235, 2005.

[20] Haider, H.K. and Ashraf, M. Bone marrow stem cell transplantation for cardiac repair. *Am. J. Physiol. Heart Circ. Physiol.*, 288:H2557–H2567, 2005.

[21] Hodgson, D.M., Behfar, A., Zingman, L.V., Kane, G.C., Perez-Terzic, C., Alekseev, A.E., Puceat, M. and Terzic, A. Stable benefit of embryonic stem cell therapy in myocardial infarction. *Am. J. Physiol. Heart Circ. Physiol.*, 287:H471–H479, 2004.

[22] Jugdutt, B.I. Remodeling of the myocardium and potential targets in the collagen degradation and synthesis pathways. *Curr. Drug Targets Cardiovasc. Haematol. Disord.*, 3:1–30, 2003.

[23] Kim, H., Kim, S.W., Nam, D., Kim, S. and Yoon, Y.S. Cell therapy with bone marrow cells for myocardial regeneration. *Antioxid. Redox. Signal*, 11:1897–1911, 2009.

[24] Kumar, D. and Jugdutt, B.I. Apoptosis and oxidants in the heart. *J. Lab. Clin. Med.*, 142:288–297, 2003.

[25] Kumar, D., Kamp, T.J. and LeWinter, M.M. Embryonic stem cells: Differentiation into cardiomyocytes and potential for heart repair and regeneration. *Coron. Artery Dis.*, 16:111–116, 2005.

[26] Kumar, D., Lou, H. and Singal, P.K. Oxidative stress and apoptosis in heart dysfunction. *Herz* 27:662-668, 2002.

[27] Lee, A.S., Tang, C., Cao, F., Xie, X., van der BK, Hwang, A., Connolly, A.J., Robbins, R.C. and Wu, J.C. Effects of cell number on teratoma formation by human embryonic stem cells. *Cell Cycle*, 8:2608–2612, 2009.

[28] Li, J.H., Zhang, N. and Wang, J.A. Improved anti-apoptotic and anti-remodeling potency of bone marrow mesenchymal stem cells by anoxic pre-conditioning in diabetic cardiomyopathy. *J. Endocrinol. Invest.*, 31:103–110, 2008.

[29] LL, E., Zhao, Y.S., Guo, X.M., Wang, C.Y., Jiang, H., Li, J., Duan, C.M. and Song, Y. Enrichment of cardiomyocytes derived from mouse embryonic stem cells. *J. Heart Lung Transplant.*, 25:664–674, 2006.

[30] Mancini, D. and Lietz, K. Selection of cardiac transplantation candidates in 2010. *Circulation*, 122:173–183, 2010.

[31] Menard, C., Hagege, A.A., Agbulut, O., Barro, M., Morichetti, M.C., Brasselet, C., Bel, A., Messas, E., Bissery, A., Bruneval, P., Desnos, M., Puceat, M. and Menasche, P. Transplantation of cardiac-committed mouse embryonic stem cells to infarcted sheep myocardium: A preclinical study. *Lancet*, 366:1005–1012, 2005.

[32] Menasche, P. Skeletal myoblasts and cardiac repair. *J. Mol. Cell Cardiol.*, 45:545–553, 2008.

[33] Min, J.Y., Yang, Y., Converso, K.L., Liu, L., Huang, Q., Morgan, J.P. and Xiao, Y.F. Transplantation of embryonic stem cells improves cardiac function in postinfarcted rats. *J. Appl. Physiol.*, 92:288–296, 2002.

[34] Monzen, K., Shiojima, I., Hiroi, Y., Kudoh, S., Oka, T., Takimoto, E., Hayashi, D., Hosoda, T., Habara-Ohkubo, A., Nakaoka, T., Fujita, T., Yazaki, Y. and Komuro, I. Bone morphogenetic proteins induce cardiomyocyte differentiation through the mitogen-activated protein kinase kinase kinase TAK1 and cardiac transcription factors Csx/Nkx-2.5 and GATA-4. *Mol. Cell Biol.*, 19:7096–7105, 1999.

[35] Nadal-Ginard, B., Anversa, P., Kajstura, J. and Leri, A. Cardiac stem cells and myocardial regeneration. *Novartis Found. Symp.*, 265:142-154, 2005.

[36] Navarro-Alvarez, N., Soto-Gutierrez, A. and Kobayashi, N. Hepatic stem cells and liver development. *Methods Mol. Biol.*, 640:181–236, 2010.

[37] Nelson, T.J., Martinez-Fernandez, A. and Terzic, A. Induced pluripotent stem cells: Developmental biology to regenerative medicine. *Nat. Rev. Cardiol.* 7:700–710, 2010.

[38] Nelson, T.J., Martinez-Fernandez, A., Yamada, S., Perez-Terzic, C., Ikeda, Y. and Terzic, A. Repair of acute myocardial infarction by human stemness factors induced pluripotent stem cells. *Circulation*, 120:408–416, 2009.

[39] Noiseux, N., Gnecchi, M., Lopez-Ilasaca, M., Zhang, L., Solomon, S.D., Deb, A., Dzau, V.J. and Pratt, R.E. Mesenchymal stem cells overexpressing Akt dramatically repair

infarcted myocardium and improve cardiac function despite infrequent cellular fusion or differentiation. *Mol. Ther.*, 14:840–850, 2006.

[40] Nussbaum, J., Minami, E., Laflamme, M.A., Virag, J.A., Ware, C.B., Masino, A., Muskheli, V., Pabon, L., Reinecke, H. and Murry, C.E. Transplantation of undifferentiated murine embryonic stem cells in the heart: Teratoma formation and immune response. *FASEB J.*, 2007.

[41] Orlic, D., Kajstura, J., Chimenti, S., Jakoniuk, I., Anderson, S.M., Li, B., Pickel, J., McKay, R., Nadal-Ginard, B., Bodine, D.M., Leri, A. and Anversa, P. Bone marrow cells regenerate infarcted myocardium. *Nature*, 410:701–705, 2001.

[42] Pan, J. and Baker, K.M. Retinoic acid and the heart. *Vitam. Horm.*, 75:257–283, 2007.

[43] Puceat, M. TGFbeta in the differentiation of embryonic stem cells. *Cardiovasc. Res.*, 2006.

[44] Reubinoff, B.E., Pera, M.F., Fong, C.Y., Trounson, A. and Bongso, A. Embryonic stem cell lines from human blastocysts: Somatic differentiation in vitro. *Nat. Biotechnol.*, 18:399–404, 2000.

[45] Schachinger, V., Assmus, B., Britten, M.B., Honold, J., Lehmann, R., Teupe, C., Abolmaali, N.D., Vogl, T.J., Hofmann, W.K., Martin, H., Dimmeler, S. and Zeiher, A.M. Transplantation of progenitor cells and regeneration enhancement in acute myocardial infarction: Final one-year results of the TOPCARE-AMI Trial. *J. Am. Coll. Cardiol.*, 44:1690–1699, 2004.

[46] Schachinger, V., Erbs, S., Elsasser, A., Haberbosch, W., Hambrecht, R., Holschermann, H., Yu, J., Corti, R., Mathey, D.G., Hamm, C.W., Suselbeck, T., Werner, N., Haase, J., Neuzner, J., Germing, A., Mark, B., Assmus, B., Tonn, T., Dimmeler, S. and Zeiher, A.M. Improved clinical outcome after intracoronary administration of bone-marrow-derived progenitor cells in acute myocardial infarction: Final 1-year results of the REPAIR-AMI trial. *Eur. Heart J.*, 27:2775–2783, 2006.

[47] Schachinger, V., Tonn, T., Dimmeler, S. and Zeiher, A.M. Bone-marrow-derived progenitor cell therapy in need of proof of concept: Design of the REPAIR-AMI trial. *Nat. Clin. Pract. Cardiovasc. Med.*, 3(Suppl 1):S23–S28, 2006.

[48] Shi, L., Mao, C., Xu, Z. and Zhang, L. Angiotensin-converting enzymes and drug discovery in cardiovascular diseases. *Drug Discov. Today*, 15:332–341, 2010.

[49] Singla, D.K. Embryonic stem cells in cardiac repair and regeneration. *Antioxid. Redox. Signal*, 11:1857–1863, 2009.

[50] Singla, D.K, Hacker, T.A., Ma, L., Douglas, P.S., Sullivan, R., Lyons, G.E. and Kamp, T.J. Transplantation of embryonic stem cells into the infarcted mouse heart: Formation of multiple cell types. *J. Mol. Cell Cardiol.*, 40:195–200, 2006.

[51] Singla, D.K., Lyons, G.E. and Kamp, T.J. Transplanted embryonic stem cells following mouse myocardial infarction inhibit apoptosis and cardiac remodeling. *Am. J. Physiol. Heart Circ. Physiol.*, 293:H1308–H1314, 2007.

[52] Singla, D.K. and McDonald, D.E. Factors released from embryonic stem cells inhibit apoptosis of H9c2 cells. *Am. J. Physiol. Heart Circ. Physiol.*, 293:H1590–H1595, 2007.

[53] Singla, D.K., Selby, D.E., Singla, R.D. and Fatma, S. Factors Released From Embryonic Stem Cells Stimulate c-kit-FlK-1+ve Progenitor Cells and Enhance Neovascularization. *Antioxid. Redox. Signal*, 2010.

[54] Singla, D.K., Singla, R.D., Lamm, S. and Glass, C. TGF{beta}2 Treatment Enhances Cytoprotective Factors Released from Embryonic Stem Cells and Inhibits Apoptosis in the Infarcted Myocardium. Am J Physiol Heart Circ Physiol 2011.

[55] Singla, D.K., Singla RD and McDonald DE. Factors released from embryonic stem cells inhibit apoptosis in H9c2 cells through PI3K/Akt but not ERK pathway. *Am. J. Physiol. Heart Circ. Physiol.*, 295:H907–H913, 2008.

[56] Singla, D.K. and Sun, B. Transforming growth factor-beta2 enhances differentiation of cardiac myocytes from embryonic stem cells. *Biochem. Biophys. Res. Commun.*, 332:135–141, 2005.

[57] Swijnenburg, R.J., Tanaka, M., Vogel, H., Baker, J., Kofidis, T., Gunawan, F., Lebl, D.R., Caffarelli, A.D., de Bruin, J,L,, Fedoseyeva, E.V. and Robbins, R.C. Embryonic stem cell immunogenicity increases upon differentiation after transplantation into ischemic myocardium. *Circulation* 112:I166–I172, 2005.

[58] Swynghedauw, B. Molecular mechanisms of myocardial remodeling. *Physiol. Rev.* 79:215–262, 1999.

[59] Taylor, F., Ward, K., Moore, T.H., Burke, M., Davey, S.G., Casas, J.P. and Ebrahim, S. Statins for the primary prevention of cardiovascular disease. Cochrane Database Syst Rev CD004816, 2011.

[60] Tendera, M., Wojakowski, W., Ruzyllo, W., Chojnowska, L., Kepka, C., Tracz, W., Musialek, P., Piwowarska, W., Nessler, J., Buszman, P., Grajek, S., Breborowicz, P., Majka, M. and Ratajczak, M.Z. Intracoronary infusion of bone marrow-derived selected CD34+CXCR4+ cells and non-selected mononuclear cells in patients with acute STEMI and reduced left ventricular ejection fraction: Results of randomized, multicentre Myocardial Regeneration by Intracoronary Infusion of Selected Population of Stem Cells in Acute Myocardial Infarction (REGENT) Trial. *Eur. Heart J.*, 30:1313–1321, 2009.

[61] Teramura, T., Onodera, Y., Murakami, H., Ito, S., Mihara, T., Takehara, T., Kato, H., Mitani, T., Anzai, M., Matsumoto, K., Saeki, K., Fukuda, K., Sagawa, N. and Osoi, Y. Mouse androgenetic embryonic stem cells differentiated to multiple cell lineages in three embryonic germ layers in vitro. *J. Reprod. Dev.*, 55:283–292, 2009.

[62] Wolf, D.P., Kuo, H.C., Pau, K.Y. and Lester, L. Progress with nonhuman primate embryonic stem cells. *Biol. Reprod.*, 71:1766–1771, 2004.

[63] Wollert, K.C., Meyer, G.P., Lotz, J., Ringes-Lichtenberg, S., Lippolt, P., Breidenbach, C., Fichtner, S., Korte, T., Hornig, B., Messinger, D., Arseniev, L., Hertenstein, B., Ganser, A. and Drexler, H. Intracoronary autologous bone-marrow cell transfer after myocardial infarction: The BOOST randomised controlled clinical trial. *Lancet*, 364:141–148, 2004.

[64] Zhang, J., Huang, C., Wu, P., Yang, J., Song, T., Chen, Y., Fan, X. and Wang, T. Differentiation induction of cardiac c-kit positive cells from rat heart into sinus node-like cells by 5-azacytidine. *Tissue Cell*, 43:67–74, 2011.

17

Potential Confounds in Stem Cell Therapy: A Case Study of Mesenchymal Stem Cells and Cancer Stem Cells

Jessian L. Munoz, Shyam A. Patel, Phillip K. Lim, Agata Giec, Kimberly A. Silverio, Lillian F. Pliner and Pranela Rameshwar

UMDNJ-New Jersey Medical School, Newark, New Jersey, USA;
e-mail: rameshwa@umdnj.edu

Abstract

Mesenchymal stem cells (MSCs) show promise for the treatment for various clinical disorders. Despite this promise there are concerns of safety. This review will outline these issues since the identification of concerns could lead to effective treatment. Unlike embryonic stem cells, MSCs are desirable stem cells for translation, mostly due to reduced ethical concerns, ease in expansion, cellular plasticity, ability to be transplanted as off-the-shelf stem cells, and reduced ability to be transformed. Indications of MSCs include cellular therapy for immune suppression such as graft versus host disease, allergies, tissue repair and protection such as traumatic brain injury and spinal cord injury. MSCs express receptors for chemokines, indicating that these cells could migrate towards tissue insults. This property makes MSCs ideal for gene and drug delivery cells. Despite the therapeutic promise for MSCs as well as robust characterization, it is difficult to accurately determine the phenotype of MSCs. This has led scientists to propose heterogeneity among MSCs. However, the seeming heterogeneity could be due to varied culture methods. This review discusses two major concerns of for the safe delivery of MSCs. Firstly, we addressed the issue of MSCs as support of tumor growth. Secondly, the review discussed the instability that could occur when MSCs

P. Di Nardo (Ed.), Adult Stem Cell Standardization, 223–237.

are placed within a milieu of inflammatory mediators. This could initiate transformation in the implanted MSCs.

Keywords: breast cancer, microenvironment, cancer stem cells, mesenchymal stem cells, exosomes.

17.1 Introduction

The recurrence of breast cancer in patients after >5 years of disease survival indicates that patients could have dormant breast cancer cells even after aggressive treatment. Others have reported on disease-free periods as long as 20–25 years [1]. In a significant number of BC recurrences, the clinical evidence has identified bone marrow as a source of the originating breast cancer cells [2–4], supporting the existence of dormant cancer cells in bone marrow. The report indicated that breast cancer can take >10 years for clinical detection [5]. This information, combined with the time of recurrence and the implication of bone marrow as the source of tertiary metastasis suggest that dissemination of breast cancer cells could home to bone marrow where the cancer cells adapt cell cycle quiescence, entering a G0-G1 arrest. During this time patients are expected to be clinically asymptomatic. The question is how this state of quiescence will impact stem cell treatment. This review discusses how this could cause untoward effects in stem cell treatment, especially since the dormant cells are resistant to chemotherapy, which mostly targets actively proliferating cells [1]. Although this review focuses on breast cancer, the discussion can be extrapolated to other cancers.

17.2 Mesenchymal Stem Cells (MSCs) and Breast Cancer

The role of MSCs needs to be highlighted with regard to dormancy of breast cancer cells. MSCs are pluripotent cells that can be found in adult and fetal tissues [6]. In bone marrow, MSCs represent a very low percentage of cells that show lineage differentiation to generate specialized cells such as stroma [7]. In human bone marrow, MSCs are found in the abluminal region of blood vessels and in contact with trabeculae [8, 9]. The presence of MSCs around the main blood vessels, combined with the immune plastic functions of MSCs, suggests the role of "gate keepers" of the bone marrow [10]. This designation is warranted when one considers that MSCs are among the first cells to be met by any cell, including cancer cells, entering bone marrow.

The role of MSCs to the breast cancer cells entering into bone marrow underscores the future of stem cell therapy. As such, it is necessary to briefly discuss how MSCs are involved in protecting as well as supporting the entering breast cancer cells. Breast cancer cells interact with MSCs via CXCL12-CXCR4 [11, 12]. This initial interaction with breast cancer cells can prompt the MSCs to inhibit natural killer activity and cytotoxic responses to protect the cancer cells from immune clearance [13]. Once in bone marrow, the CXCR4-expressing breast cancer cells can be attracted by the region where the ligand, CXCL12, is high, which is close to the endosteum [14]. At the region close to the endosteum, breast cancer cells can attain quiescence through the establishment of gap junctional intercellular communication (GJIC) with the endogenous stroma [15]. This microenvironment is favorable for the cancer cells because the cells can take advantage of the method that protects the endogenous hematopoietic stem cells.

In addition to creating a dormant microenvironment for breast cancer cells, MSCs have the ability to recruit and be recruited by tumor cells via the release of cytokines and growth factors [7, 16]. Similar to areas of inflammation and tissue repair, extracellular matrix (ECM) proteases can attract MSCs to the microenvironment surrounding the tumor [10]. As an example, activation of the urokinase plasminogen activator (uPA) and the urokinase plasminogen activator receptor (uPAR), which are upregualated in solid tumors, are believed to be correlated with increased MSC migration [17]. In addition, chemokines, present at the tumor areas can also attract MSCs to the site of tumors [18]. Similarly, CXCR4-expressing tumors can be attracted to activated MSCs. Once in the microenvironment of a primary or secondary tumor, MSCs can form tumor-associated fibroblasts, pericytes, and myofibroblasts and release cytokines, such as vascular endothelial growth factor (VEGF) that support growth and survival [19].

The ability of MSCs to be attracted to areas of tumors has led to proposals that scientists can take advantage of this property to deliver drugs and genes to the area, then subject the MSCs to undergo cell death by transfecting with a suicide gene [20]. The use of MSCs as cellular vehicle for drug delivery is desirable, mostly due to the ease of isolating these stem cells from volunteers and the ability to expand from discarded tissue such as adipose tissues. MSCs seems to be attractive drug delivery for the treatment in cancers such as glioblastoma, with poor prognosis, in addition to difficulty to deliver of chemotherapy, posed by the blood brain barrier [21]. Indeed, Nakamizo et al. [21] reported on the localization of MSCs, engineered with anti-glioma agent, and what this can do to the tumor. It is important to understand the crosstalk between MSCs and their microenvironment for safe delivery of these stem cells to patients since MSCs are important in cancer metastasis as well as protection.

17.3 Spontaneous Transformation of MSCs

This section discusses the possibility that MSCs could be transformed in culture as well as *in vivo*. This issue is fundamental to the safe delivery of MSCs, as well as other stem cells since the expression of cytokine receptors will cause the implanted stem cells to establish rapid crosstalk with the tissue microenvironment.

Direct transformation of MSCs is a controversial topic. In 2005, Rubio and colleagues reported that adipose-derived MSCs, in culture for long periods, could spontaneously generate tumors in immune compromised mice [22, 23]. It was observed that MSCs reached senescence after two months and then spontaneously overcame this phase, which was followed with increased proliferation and trisomy of chromosome 8. Cell cycle analysis showed chromosomal instability and loss of pluripotency. Cells lost general spindle phenotype and cell surface markers (CD34/90/105). Yet, it was later determined by DNA fingerprinting and Short Tandem Repeat analysis, that the cell lines had been contaminated by fibrosarcoma, osteosarcoma and glioblastoma cell lines.

In a later study, in 2006, using murine bone marrow-derived MSCs, Miura and colleagues reported spontaneous transformation with associated tumor formation in immunocompromised grafted mice after 29 passages *in vitro* [24]. Passage 29 cells lost osteogenic differentiation capacities, increased proliferative capacities and were morphologically small and round. The transformation and proliferative capacities correlated with the number of passages as the passage number reached number 92. At passage 55, 90% of analyzed MSC clones showed chromosomal imbalance. Histological analyses of the generated tumors presented with positive staining for vimentin and negative staining for cytokeratin, leading to the determination of MSC-derived fibrosarcomas. Passage 65 MSCs produced tumors in multiple locations following tail-vein injection, including the lungs, kidneys and vertebrae. Taken together, the data concluded that murine MSCs, if passaged for long periods can acquire genetic imbalances, with loss of pluripotency and gain of tumorogenicity.

It was later, in 2009, that the controversy lying behind spontaneous transformation of human MSCs was finally solved. Rosland et al. showed that a subset of MSCs (~46%), after long-term culture, exhibited malignant oncogenic formation [25]. These MSCs were renamed as Transformed Mesenchymal Cells (TMC); the TMCs showed loss of differentiation capacities, increased proliferation, altered morphology and growth on soft agar assays. TMCs were also highly malignant, after intra-venous delivery into immune compromised mice. The TMCs rapidly (9 days) colonized the lungs. The mice, euthanized after 39 days, showed evidence of cancer burden.

Due to the growing evidence of MSC based treatments, the stability of these bone marrow-derived cells is of considerable importance. Although oncogenic capacities will have to be analyzed before cellular treatment for human disease, MSC-derived tumors may serve as a model for anti-neoplastic pharmacological treatment and radiotherapy. In addition, the origins of many human malignancies may lie in MSC transformation and a better understanding of the transformation process may lead to new and enhanced treatments or prophylaxis.

17.3.1 MSC Origin of Cancers

Although the experimental evidence shows spontaneous transformation of MSCs *in vitro*, a number of human malignancies may occur by *in vivo* transformations of localized and recruited MSCs. These include gastric cancer and Ewing's tumors. Ewing's sarcoma (EWS) is a rare tumor of the bone, mostly in children and young adults. This poorly differentiated tumor shows great metastatic potential, commonly colonizing the lungs and bone marrow. The initiated molecular mechanisms are caused by a fusion of the EWS gene and other family members. The origin of EWS is highly debatable, but there is evidence that EWS may be of primitive neuroectodermal origin. On the other hand, MSCs have also been proposed as the origin. The fact the MSC may also be of neuroectodermal origin reinforces this concept. In order to show evidence of MSCs as the origin of EWS, the gene was knocked out and this resulted in the ability of the cells to differeniate into chondrocytes, adipocytes and osteoblasts, consistent with the lineage potential of MSCs. Also, if the EWS gene fusion is inserted into murine MSCs, MCSs transform into Ewing's sarcomas, although similar findings were not observed in human MSCs [26, 27].

Inflammation and *H. pylori* colonization have been attributed to the initiating cause of gastric cancer, specifically glandular gastric adenocarcioma. MSCs are known to migrate to sites of inflammation. Growth signals and secreted cytokines at sites of injury and inflammation such as *H. pylori* infection may affect the pluripotent state of MSCs, leading to transformation [28]. In a mouse model of gastric cancer, followed by bone marrow transplantation with beta-galactosidase (beta gal) cells from donor mice, Houghton and colleagues showed migration of bone marrow components 20 weeks after initial infection. Dysplastic glands were shown to be of bone marrow origin by beta-gal expression. Uncovering the origins of tumors is of great significance. If MSCs can generate tumors *in vivo*, then the microenvironment of transplanted MSCs for any treatment would be significant to cellular transformation. In addition, prophylactic treatment of tumor development may target these transforming events.

17.4 MSC Transformation: Role of the Microenvironment

Stem cells maintain a state of cellular plasticity, this plasticity may be greatly altered by the microenvironment in which these cells are placed; MSCs are no exception. If stem cells are planned for therapy, their placement within a milieu of inflammatory mediators is significant for consideration whether the microenvironment can affect the stem cells. This is particularly true because several of the genes associated with pluripotency are also linked to oncogenesis as well as tumor suppression.

Repressor Element-1 Silencing Transcription factor (REST), also known as Neuron Restrictive Silencing Factor (NRSF), is a DNA-binding protein that exerts both tumor suppressor and oncogenic properties [29]. MSCs express high levels of REST protein [30]. Microenvironmental chemokines and inflammatory molecules can affect REST expression and lead to MSC differentiation and possible transformation. Sites of tissue injuries are complex with the presence of multiple soluble and insoluble mediators as well as varied immune cell subsets. Among these, IL-1 may exert direct effects and may regulate other cytokines with positive and negative effects.

IL-1α has been shown to cause a rapid decrease in *REST* expression in MSCs. While this increase could be an advantage to tissue repair, the rapid decrease in *REST* expression could predispose the cell to transformation. IL-1 stimulation of MSCs or the early-differentiated MSCs towards neurogenesis causes a rapid decrease in REST with concomitant increase in other genes, such as *TAC1*, shown to exert oncogenic functions [31–33].

17.5 MSCs Role in Tumor-Initiating Cells

The field of cancer stem cells is not mutually exclusive of considerations on safe transplantation of MSCs. This section discusses possible unintended effects of MSCs on cancer stem cells. Although the literature on the cancer stem cell field has exploded in recent years, the model was first proposed more than five decades ago. The first definitive evidence for a stem cell basis for cancer was the discovery by Bonnet and Dick in 1997 that only particular cells or subsets within the hematopoietic hierarchy were capable of *in vivo* repopulation of leukemia [34]. The authors reported that leukemia arose from primitive hematopoietic cells, namely cells of the CD34+/CD38- phenotype [34]. Over the past decade, scientists have fine-tuned the theory with the identification of specific markers for specific cell types.

In order to understand the concept of the cancer stem cell, it is important to consider the traditional theories on oncogenesis. Originally, scientists believed that cancer could be explained by a stochastic model, in which each cancer cell was thought to have tumorigenic potential [35]. However, in the light of accumulating evidence indicating heterogeneity among tumors, this model has been replaced by the hierarchical model, in which only particular subsets of cancer cells harbor tumorigenic capability. The hierarchical model holds that stem-like cells give rise to all elements of a tumor, including cells that self-renew and cells that proceed to differentiate into mature tissue types [35]. Thus, both in vitro findings and clinical findings have suggested that tumors consist of heterogeneous progeny, including stem cells, progenitors, transient-amplifying cells, and differentiated cells [36].

Molecular evidence on the stem cell basis for cancer has arisen from the identification of a variety of cell surface markers and intracellular signaling pathways that are common to both stem cells and cancer cells. For instance, regarding cell surface markers, CD133 (prominin-1) has been found to be expressed in glioma tumor-initiating stem cells [37]. As another example, the CD117 (c-kit) phenotype has been found in stem-like cells in osteosarcoma [38]. With regards to intracellular signal transduction pathways, Wnt and Notch signaling are dysregulated in numerous cancers, and these molecules have roles in stem cell biology and embryological development [39]. The institution of targeted therapy may be challenged by the fact that resident stem cells in normal tissue can be harmed in process of attempting to eradicate the cancer stem cells. Therapeutic avenues that have been under consideration include oncolytic viral therapy, immunotoxin therapy, differentiation therapy, microenvironmental manipulation, pharmacological disruption, and nanotechnology [40–42]. A detailed discussion of each of these modalities is beyond the scope of this chapter, but the prospects remain promising.

Based on the rapidly evolving data, scientists are becoming increasingly willing to accept that cancer is a disease of stem cells. In addition to the idea that stem cells are the source of tumors, one of the premises of the cancer stem cell hypothesis is that a stem-like subset is responsible for metastasizing to distant sites and establishing dormancy in target organs [43]. A common site of distant metastasis for cancer is the bone marrow, with resident MSCs. As discussed above, MSCs are unique in that they have been known to exert pan-inhibitory effects on the immune system, such as suppression of T cell proliferation and natural killer cell function [44, 45]. The implications of immunosuppressive functions by MSCs are significant for treating inflammatory disorders such as autoimmune conditions. However, the immunosuppressive properties have implications for cancer progression due to the anti-tumor properties. For example, recent studies demonstrated that MSCs inhibit cytotoxic T cell and natural killer cell function, preventing the destruction of breast cancer cells [46]. Furthermore, MSCs have been shown to induce the regulatory T cell phenotype, which has implications in dormancy in the tumor microenvironment [46, 47].

In contrast to immune protection by MSCs, recent experimental evidence suggested a bimodal role for MSCs on cancer progression [48, 49]. Hence, the MSCs can act as a double-edged sword. MSCs have also been shown to suppress breast cancer progression via ectopic expression of anti-viral interferons [48]. The role of MSCs on breast cancer progression does not always involve the stemness, but could involve the differentiated cells. For example, stromal cells, differentiated from MSCs, can form GJIC with breast cancer cells [49]. Although the role of stroma on breast cancer survival is under investigation, the evidence indicates that soluble factors such as CXCL12 could be a mediator [50]. In addition, soluble factors produced from tumors can change the functions of MSCs to carcinoma-associated fibroblast to facilitate the progression of cancer [50].

Thus far, the majority of studies involving MSCs and tumor cell interaction focus on heterogeneous cancer cells. The role of particular subsets in the establishment of dormancy in bone marrow and other organs remains to be investigated. It is this uncertainty that makes it difficult for the translation of MSCs to patients. Current speculation holds that stem-like cancer cells may have preferential role in this phase. To date, the data on the mechanism, which we propose to be complex, is unclear, although the area remains open to investigation.

17.6 Exosomes

Any review on the safety of stem cells needs to include exosomes due to the role in facilitating intercellular communication. Such interactions are becoming increasingly more important in order to understand biological processes ranging from stimulation of immune cells to cancer and stem cell biology [49, 51]. Although intercellular communications can occur by multiple methods, exosomes are becoming a key area of research.

Exosomes are small (50–90 nm) membrane vessicles that are released into the extracellular environment when intracellular multivesicular bodies fuse with the plasma membrane [52]. Exosomes were first identified in studies on the process of transferrin receptor release through small vesicles in maturing reticulocytes [53]. Since then, exosomes have been isolated from the cell culture supernatant of several hematopoietic and non-hematopoietic cells, including mature immune cells, intestinal epithelial cells, neuronal cells, tumor cells and human embryonic stem cells derived-mesenchymal stem cells [54–61]. Just as the list of cell types that utilize exosomes is not all inclusive, their exact functions have all yet to be elucidated.

Interest in exosome function was sparked with the discovery that B-cells and dendritic cells release exosomes containing MHC class II and thus, play a role in antigen presentation [55, 60]. Furthermore, the exosomes produced by dendritic cells have been shown to promote anti-tumor response in vivo by priming cytotoxic T lymphocytes ag*ainst e*stablished murine tumors [60]. The ability of exosomes, secreted from antigen presenting cells, to participate in [60] antigen presentation and activate specific cell types is in part due to their composition, which includes the presence of MHC class I and II, adhesion molecules, cytosolic chaperone proteins, cytoskeletal proteins, and other enzymes [62]. Additionally, the presence of tetraspanins such as CD9, CD 63, CD81, and CD83, and proteins that participate in vesicle formation and trafficking, e.g. Alix, is characteristic of exosomes and often used as identification markers [62, 63]. The ability to isolate and identify exosomes has been crucial in identifying other functions of exosomes, including their potential use as a biomarker. For example, the presence of exosomes containing aquaporin-2 protein in urine may aid in the diagnosis of diabetes insipidus in patients [64] while the presence of psoriasin in exosomes makes these vessicles an attractive biomarker for squamous cell bladder cancer [65]. By searching for the presence of exosomes and analyses on their composition, these studies may provide a new tool in varied aspects of medicine and the translation of stem cell to patients.

In addition to their ability to transport proteins, exosomes have been implicated in the exchange of genetic material among cells. More specifically, exosomes in murine and human mast cells have been shown to contain RNA, such as translatable messenger RNA (mRNA) and functional microRNA (miRNA) [66]. The identification of RNA within exosomes provides a novel mechanism by which a cell could deliver RNA to another cell and modulate recipient-cell protein expression. The presence of miRNAs is of significant interest because this indicates cleavage of mRNA and/or inhibition of translation. The existence of small regulatory RNA within exosomes and their ability to interact with specific cell-types reveals a potential use for exosomes as targeted therapy, including cancers.

It has been shown that certain cancers can attain a state of quiescence and evade detection by immune cells or resist chemotherapeutics. One example of this is the ability of metastatic breast cancer cells to enter into a dormant state within the bone marrow [12]. Because of their low level of activity and replication, these dormant cancer cells are not suitable candidates for traditional mitostatic treatments and are thus fairly difficult to treat. However, there is hope in using exosomes to target these cancer cells because of their ability to target specific cell-types and the presence of miRNA within the cancer cells. Deep sequencing analysis has revealed that MSCs from embryonic stem cells can release exosomes that contain the let-7 family of miRNA transcripts and regulate cellular differentiation [61]. Other studies have shown the MSCs can reduce infarct size in ischemic myocardial tissue through the secretion of exosomes, cytokines, and growth factors [67]. These results bring up the question whether exosome production by transplanted MSCs is possible, and if so, what the effects could be.

Continued advances in bioinformatics and sequencing technology will allow for deeper investigation into the regulatory mechanisms involved in stem cell and cancer dormancy and may identify potential targets for miRNA found within exosomes. In conjunction, more studies with MSCs must be completed in order to determine their putative ability to secrete functional exosomes. Over time, however, it may be possible to forego the necessity to have a cell produce exosomes and instead, synthesize them for cell-less therapy. As additional information becomes available about the composition of exosomes and their functions, they may develop utility, not only as biomarkers but also as targeted cancer therapy.

17.7 Conclusion

This review highlights several ongoing issues, with the translation of MSCs, as well as other stem cells to patients. It is difficult to discuss the stem cell treatment without including the field of cancer biology as well as the immunology of stem cells. Although these two fields appear to be different, they are not mutually exclusive, with regard to the response of stem cells at sites of tissue injury. There are several evolving fields such as miRNAs and exosomes that are included in the discussion. In addition to inflammatory mediators, stem cells can produce exosomes to establish intracellular crosstalk with other cells. Thus, although we discussed the possibility of stem cells supporting pre-existing cancers, the induction of cancers via exosome release cannot be discounted. Overall, this review proposes teams of multi-disciplinary scientists, as well as clinical and basic scientists, to collaborate for safe delivery of stem cells to patients.

References

[1] J.A. Aguirre-Ghiso, Models, mechanisms and clinical evidence for cancer dormancy, *Nat. Rev. Cancer*, 7:834–846, 2007.

[2] B. Naume, X. Zhao, M. Synnestvedt, E. Borgen, H.G. Russnes, O.C. Lingjaerde, M. Stromberg, G. Wiedswang, G. Kvalheim, R. Karesen, J.M. Nesland, A.L. Borresen-Dale, and T. Sorlie, Presence of bone marrow micrometastasis is associated with different recurrence risk within molecular subtypes of breast cancer, *Mol. Oncol.*, 1:160–171, 2007.

[3] M. Habeck, Bone-marrow analysis predicts breast-cancer recurrence, *Mol. Med. Today*, 6:256–257, 2000.

[4] S. Riethdorf, H. Wikman, and K. Pantel, Review: Biological relevance of disseminated tumor cells in cancer patients, *Int. J. Cancer*, 123:1991–2006, 2008.

[5] J.E. Talmadge, Clonal selection of metastasis within the life history of a tumor, *Cancer Res.*, 67:11471–11475, 2007.

[6] S.A. Patel, L. Sherman, J. Munoz, and P. Rameshwar, Immunological properties of mesenchymal stem cells and clinical implications, *Arch. Immunol. Ther. Exp.*, 56:1–8, 2008.

[7] L. da Silva Meirelles, A.M. Fontes, D.T. Covas, and A.I. Caplan, Mechanisms involved in the therapeutic properties of mesenchymal stem cells, *Cytokine & Growth Factor Rev.*, 20:419–427, 2010.

[8] P. Rameshwar, Potential novel targets in breast cancer, *Curr. Pharm. Biotechnol.*, 10:148–153, 2009.

[9] P. Bianco, P.G. Robey, and P.J. Simmons, Mesenchymal stem cells: Revisiting history, concepts, and assays, *Cell Stem Cell*, 2:313–319, 2008.

[10] J.L. Chan, K.C. Tang, A.P. Patel, L.M. Bonilla, N. Pierobon, N.M. Ponzio, and P. Rameshwar, Antigen-presenting property of mesenchymal stem cells occurs during a narrow window at low levels of interferon-γ, *Blood*, 107:4817–4824, 2006.

[11] S.A. Patel, A.C. Heinrich, B.Y. Reddy, B. Srinivas, N. Heidaran, and P. Rameshwar, Breast cancer biology: The multifaceted roles of mesenchymal stem cells, *J Oncol.*, 2008:425895, 2008.

[12] K.E. Corcoran, K.A. Trzaska, H. Fernandes, M. Bryan, M. Taborga, V. Srinivas, K. Packman, P.S. Patel, and P. Rameshwar, Mesenchymal stem cells in early entry of breast cancer into bone marrow, *PLoS. One*, 3:e2563, 2008.

[13] S.A. Patel, J.R. Meyer, S.J. Greco, K.E. Corcoran, M. Bryan, and P. Rameshwar, Mesenchymal stem cells protect breast cancer cells through regulatory T cells: Role of mesenchymal stem cell-derived TGF-beta1, *J. Immunol.*, 184:5885–5894, 2010.

[14] T. Yin and L. Li, The stem cell niches in bone, *J. Clin. Invest.*, 116:1195–1201, 2006.

[15] A.L. Moharita, M. Taborga, K.E. Corcoran, M. Bryan, P.S. Patel, and P. Rameshwar, SDF-1α regulation in breast cancer cells contacting bone marrow stroma is critical for normal hematopoiesis, *Blood*, 108:3245-3252, 2006.

[16] Y. Rattigan, J.M. Hsu, P.J. Mishra, J. Glod, and D. Banerjee, Interleukin 6 mediated recruitment of mesenchymal stem cells to the hypoxic tumor milieu, *Exp. Cell Res.*, 316:3417–3424, 2010.

[17] M. Gutova, J. Najbauer, R.T. Frank, S.E. Kendall, A. Gevorgyan, M.Z. Metz, M. Guevorkian, M. Edmiston, D. Zhao, C.A. Glackin, S.U. Kim, and K.S. Aboody, Urokinase plasminogen activator and urokinase plasminogen activator receptor mediate human stem cell tropism to malignant solid tumors, *Stem Cells*, 26:1406–1413, 2008.

[18] S.A. Park, C.H. Ryu, S.M. Kim, J.Y. Lim, S.I. Park, C.H. Jeong, J.A. Jun, J.H. Oh, S.H. Park, W. Oh, and S.S. Jeun, CXCR4-transfected human umbilical cord blood-derived mesenchymal stem cells exhibit enhanced migratory capacity toward gliomas, *Int. J. Oncol.*, 38:97–103, 2011.

[19] S.A. Bergfeld and Y.A. DeClerck, Bone marrow-derived mesenchymal stem cells and the tumor microenvironment, *Cancer Metastasis Rev.*, 29:249–261, 2010.

[20] L. Kucerova, V. Altanerova, M. Matuskova, S. Tyciakova, and C. Altaner, Adipose tissue-derived human mesenchymal stem cells mediated prodrug cancer gene therapy, *Cancer Res.*, 67:6304–6313, 2007.

[21] A. Nakamizo, F. Marini, T. Amano, A. Khan, M. Studeny, J. Gumin, J. Chen, S. Hentschel, G. Vecil, J. Dembinski, M. Andreeff, and F.F. Lang, Human bone marrow-derived mesenchymal stem cells in the treatment of gliomas, *Cancer Res.*, 65:3307–3318, 2005.

[22] D. Rubio, J. Garcia-Castro, M.C. Martin, F.R. de la, J.C. Cigudosa, A.C. Lloyd, and A. Bernad, Spontaneous human adult stem cell transformation, *Cancer Res.*, 65:3035–3039, 2005.

[23] D. Rubio, S. Garcia, M.F. Paz, C.T. De la, L.A. Lopez-Fernandez, A.C. Lloyd, J. Garcia-Castro, and A. Bernad, Molecular characterization of spontaneous mesenchymal stem cell transformation, *PLoS. One.*, 3:e1398, 2008.

[24] M. Miura, Y. Miura, H.M. Padilla-Nash, A.A. Molinolo, B. Fu, V. Patel, B.M. Seo, W. Sonoyama, J.J. Zheng, C.C. Baker, W. Chen, T. Ried, and S. Shi, Accumulated chromosomal instability in murine bone marrow mesenchymal stem cells leads to malignant transformation, *Stem Cells*, 24:1095–1103, 2006.

[25] G.V. Rosland, A. Svendsen, A. Torsvik, E. Sobala, E. McCormack, H. Immervoll, J. Mysliwietz, J.C. Tonn, R. Goldbrunner, P.E. Lonning, R. Bjerkvig, and C. Schichor, Long-term cultures of bone marrow-derived human mesenchymal stem cells frequently undergo spontaneous malignant transformation, *Cancer Res.*, 69:5331–5339, 2009.

[26] P.P. Lin, Y. Wang, and G. Lozano, Mesenchymal stem cells and the origin of Ewing's sarcoma, *Sarcoma*, 2011.

[27] F. Tirode, K. Laud-Duval, A. Prieur, B. Delorme, P. Charbord, and O. Delattre, Mesenchymal stem cell features of Ewing tumors, *Cancer Cell*, 11:421–429, 2007.

[28] J. Houghton, C. Stoicov, S. Nomura, A.B. Rogers, J. Carlson, H. Li, X. Cai, J.G. Fox, J.R. Goldenring, and T.C. Wang, Gastric cancer originating from bone marrow-derived cells, *Science*, 306:1568–1571, 2004.

[29] B.Y. Reddy, S.J. Greco, P.S. Patel, K.A. Trzaska, and P. Rameshwar, RE-1-silencing transcription factor shows tumor-suppressor functions and negatively regulates the oncogenic TAC1 in breast cancer cells, *Proc. Natl. Acad. Sci.*, 106:4408–4413, 2009.

[30] K.A. Trzaska, B.Y. Reddy, J.L. Munoz, K.Y. Li, J.H. Ye, and P. Rameshwar, Loss of RE-1 silencing factor in mesenchymal stem cell-derived dopamine progenitors induces functional maturity, *Mol. Cell Neurosci.*, 39:285–290, 2008.

[31] S.J. Greco, S.V. Smirnov, R.G. Murthy, and P. Rameshwar, Synergy between the RE-1 silencer of transcription and NFkappaB in the repression of the neurotransmitter gene TAC1 in human mesenchymal stem cells, *J. Biol. Chem.*, 282:30039–30050, 2007.

[32] A.S. Singh, A. Caplan, K.E. Corcoran, J.S. Fernandez, M. Preziosi, and P. Rameshwar, Oncogenic and metastatic properties of preprotachykinin-I and neurokinin-1 genes, *Vas. Pharmacol.*, 45:235–242, 2006.

[33] D. Singh, D.D. Joshi, M. Hameed, J. Qian, P. Gascon, P. B. Maloof, A. Mosenthal, and P. Rameshwar, Increased expression of preprotachykinin-I and neurokinin receptors in human breast cancer cells: Implications for bone marrow metastasis, *Proc. Natl. Acad. Sci.*, 97:388–393, 2000.

[34] D. Bonnet and J.E. Dick, Human acute myeloid leukemia is organized as a hierarchy that originates from a primitive hematopoietic cell, *Nat. Med.*, 3:730–737, 1997.

[35] A. Sottoriva, J.J. Verhoeff, T. Borovski, S.K. McWeeney, L. Naumov, J.P. Medema, P.M. Sloot, and L. Vermeulen, Cancer stem cell tumor model reveals invasive morphology and increased phenotypical heterogeneity, *Cancer Res.*, 70:46–56, 2010.

[36] I.A. Di, F. Grizzi, C. Sherif, C. Matula, and M. Tschabitscher, Angioarchitectural heterogeneity in human glioblastoma multiforme: A fractal-based histopathological assessment, *Microvasc. Res.*, 2010.

[37] F. Jin, C. Gao, L. Zhao, H. Zhang, H.T. Wang, T. Shao, S.L. Zhang, Y.J. Wei, X.B. Jiang, Y.P. Zhou, and H.Y. Zhao, Using CD133 positive U251 glioblastoma stem cells to establish nude mice model of transplanted tumor, *Brain Res.*, 1368:82–90, 2011.

[38] A.S. Adhikari, N. Agarwal, B.M. Wood, C. Porretta, B. Ruiz, R.R. Pochampally, and T. Iwakuma, CD117 and Stro-1 identify osteosarcoma tumor-initiating cells associated with metastasis and drug resistance, *Cancer Res.*, 70:4602–4612, 2010.

[39] J. Waaler, O. Machon, J.P. von Kries, S.R. Wilson, E. Lundenes, D. Wedlich, D. Gradl, J.E. Paulsen, O. Machonova, J.L. Dembinski, H. Dinh, and S. Krauss, Novel synthetic antagonists of canonical Wnt signaling inhibit colorectal cancer cell growth, *Cancer Res.*, 71:197–205, 2011.

[40] L. Ahtiainen, C. Mirantes, T. Jahkola, S. Escutenaire, I. Diaconu, P. Osterlund, A. Kanerva, V. Cerullo, and A. Hemminki, Defects in innate immunity render breast cancer initiating cells permissive to oncolytic adenovirus, *PLoS. One.*, 5:e13859, 2011.

[41] E.I. Galanzha, J.W. Kim, and V.P. Zharov, Nanotechnology-based molecular photoacoustic and photothermal flow cytometry platform for in-vivo detection and killing of circulating cancer stem cells, *J. Biophotonics.*, 2:725–735, 2009.

[42] R. Roy, P. Willan, R. Clarke, and G. Farnie, Differentiation therapy: targeting breast cancer stem cells to reduce resistance to radiotherapy and chemotherapy, *Breast Cancer Res.*, 12(Suppl 1):O5, 2010.

[43] T. Hiraga, S. Ito, and H. Nakamura, Side population in MDA-MB-231 human breast cancer cells exhibits cancer stem cell-like properties without higher bone-metastatic potential, *Oncol. Rep.*, 25:289–296, 2011.

[44] G.M. Spaggiari, A. Capobianco, H. Abdelrazik, F. Becchetti, M.C. Mingari, and L. Moretta, Mesenchymal stem cells inhibit natural killer-cell proliferation, cytotoxicity, and cytokine production: Role of indoleamine 2,3-dioxygenase and prostaglandin E2, *Blood*, 111:1327–1333, 2008.

[45] J.A. Potian, H. Aviv, N.M. Ponzio, J.S. Harrison, and P. Rameshwar, Veto-like activity of mesenchymal stem cells: functional discrimination between cellular responses to alloantigens and recall antigens, *J. Immunol.*, 171:3426–3434, 2003.

[46] S.A. Patel, J.R. Meyer, S.J. Greco, K.E. Corcoran, M. Bryan, and P. Rameshwar, Mesenchymal stem cells protect breast cancer cells through regulatory T cells: Role of mesenchymal stem cell-derived TGF-beta, *J. Immunol.*, 184:5885–5894, 2010.

[47] Y. Wang, A. Zhang, Z. Ye, H. Xie, and S. Zheng, Bone marrow-derived mesenchymal stem cells inhibit acute rejection of rat liver allografts in association with regulatory T-cell expansion, *Transplant. Proc.*, 41:4352–4356, 2009.

[48] X. Ling, F. Marini, M. Konopleva, W. Schober, Y. Shi, J. Burks, K. Clise-Dwyer, R.Y. Wang, W. Zhang, X. Yuan, H. Lu, L. Caldwell, and M. Andreeff, Mesenchymal stem cells overexpressing IFN-beta inhibit breast cancer growth and metastases through Stat3 signaling in a syngeneic tumor model, *Cancer Microenviron.*, 3:83–95, 2010.

[49] P.K. Lim, S.K. Bliss, S.A. Patel, M. Taborga, M. Dave, L.A. Gregory, S.J. Greco, M. Bryan, P.S. Patel, and P. Rameshwar, Gap junction-mediated import of microRNA from bone marrow stromal cells can elicit cell cycle quiescence in breast cancer cells, *Cancer Res.*, 2011, in press.

[50] P.J. Mishra, P.J. Mishra, R. Humeniuk, D.J. Medina, G. Alexe, J.P. Mesirov, S. Ganesan, J.W. Glod, and D. Banerjee, Carcinoma-associated fibroblast-like differentiation of human mesenchymal stem cells, *Cancer Res.*, 68:4331–4339, 2008.

[51] S.H. Ramkissoon, P.S. Patel, M. Taborga, and P. Rameshwar, Nuclear factor-kB is central to the expression of truncated neurokinin-1 receptor in breast cancer: Implication for breast cancer cell quiescence within bone marrow stroma, *Cancer Res.*, 67:1653–1659, 2007.

[52] G. Van Niel, I. Porto-Carreiro, S. Simoes, and G. Raposo, Exosomes: A common pathway for a specialized function, *J. Biochem.*, 140:13–21, 2006.

[53] R.M. Johnstone, M. Adam, J.R. Hammond, L. Orr, and C. Turbide, Vesicle formation during reticulocyte maturation. Association of plasma membrane activities with released vesicles (exosomes), *J Biol Chem.*, 262:9412–9420, 1987.

[54] G. Raposo, D. Tenza, S. Mecheri, R. Peronet, C. Bonnerot, and C. Desaymard, Accumulation of major histocompatibility complex class II molecules in mast cell secretory granules and their release upon degranulation, *Mol. Biol. Cell*, 8:2631–2645, 1997.

[55] G. Raposo, H.W. Nijman, W. Stoorvogel, R. Liejendekker, C.V. Harding, C.J. Melief, and H.J. Geuze, B lymphocytes secrete antigen-presenting vesicles, *J. Exp. Med.*, 183:1161–1172, 1996.

[56] N. Blanchard, D. Lankar, F. Faure, A. Regnault, C. Dumont, G. Raposo, and C. Hivroz, TCR activation of human T cells induces the production of exosomes bearing the TCR/CD3 complex, *J. Immunol.*, 168:3235–3241, 2002.

[57] G. Van Niel, G. Raposo, C. Candalh, M. Boussac, R. Hershberg, N. Cerf-Bensussan, and M. Heyman, Intestinal epithelial cells secrete exosome-like vesicles, *Gastroenterology*, 121:337–349, 2001.

[58] J. Faure, G. Lachenal, M. Court, J. Hirrlinger, C. Chatellard-Causse, B. Blot, J. Grange, G. Schoehn, Y. Goldberg, V. Boyer, F. Kirchhoff, G. Raposo, J. Garin, and R. Sadoul, Exosomes are released by cultured cortical neurones, *Mol. Cell Neurosci.*, 31:642–648, 2006.

[59] J. Wolfers, A. Lozier, G. Raposo, A. Regnault, C. Thery, C. Masurier, C. Flament, S. Pouzieux, F. Faure, T. Tursz, E. Angevin, S. Amigorena, and L. Zitvogel, Tumor-derived exosomes are a source of shared tumor rejection antigens for CTL cross-priming, *Nat. Med.*, 7:297–303, 2001.

[60] L. Zitvogel, A. Regnault, A. Lozier, J. Wolfers, C. Flament, D. Tenza, P. Ricciardi-Castagnoli, G. Raposo, and S. Amigorena, Eradication of established murine tumors using a novel cell-free vaccine: Dendritic cell-derived exosomes, *Nat. Med.*, 4:594–600, 1998.

[61] W. Koh, C. Sheng, B. Tan, Q. Lee, V. Kuznetsov, L. Kiang, and V. Tanavde, Analysis of deep sequencing microRNA expression profile from human embryonic stem cells derived mesenchymal stem cells reveals possible role of let-7 microRNA family in downstream targeting of Hepatic Nuclear Factor 4 Alpha, *BMC Genomics*, 11:S6, 2010.

[62] B. Fevrier and G. Raposo, Exosomes: Endosomal-derived vesicles shipping extracellular messages, *Curr. Opin. Cell Biol.*, 16:415–421, 2004.

[63] J.S. Schorey and S. Bhatnagar, Exosome function: from tumor immunology to pathogen biology, *Traffic.*, 9:871–881, 2008.

[64] T. Pisitkun, R.F. Shen, and M.A. Knepper, Identification and proteomic profiling of exosomes in human urine, *Proc. Natl. Acad. Sci.*, 101:13368–13373, 2004.

[65] T. Pisitkun, R. Johnstone, and M.A. Knepper, Discovery of urinary biomarkers, *Mol. & Cell Proteomics*, 5:1760–1771, 2006.

[66] H. Valadi, K. Ekstrom, A. Bossios, M. Sjostrand, J.J. Lee, and J.O. Lotvall, Exosome-mediated transfer of mRNAs and microRNAs is a novel mechanism of genetic exchange between cells, *Nat. Cell Biol.*, 9:654–659, 2007.

[67] R.C. Lai, F. Arslan, M.M. Lee, N.S. Sze, A. Choo, T. S. Chen, M. Salto-Tellez, L. Timmers, C.N. Lee, R.M. El Oakley, G. Pasterkamp, D.P. de Kleijn, and S.K. Lim, Exosome secreted by MSC reduces myocardial ischemia/reperfusion injury, *Stem Cell Res.*, 4:214–222, 2010.

Author Index

Subject Index

About the Editor

Paolo Di Nardo, MD, is an experimental and clinical cardiologist and directs the Laboratory of Cellular and Molecular Cardiology at the University of Rome Tor Vergata. He is founder of the Japanese-Italian and Canadian-Italian Tissue Engineering Laboratories and a member of the Board of the National Institute for Cardiovascular Research (INRC), Italy. He acts as scientific advisor of major international organizations and member of international scientific societies, and is the author of several papers published in peer-reviewed international scientific journals and editor of various books. In 1994, he organized the first international congress in which the possibility of heart regeneration in mammalians was analyzed. Since then, his major interests have been in stem cell and tissue engineering technology.